REACHABLE SETS OF DYNAMIC SYSTEMS

Uncertainty, Computational Techniques, and Decision Intelligence Book Series

Series Editors

Tofigh Allahviranloo, PhD
Faculty of Engineering and Natural Sciences, Istinye University, Istanbul, Turkey

Narsis A. Kiani, PhD
Algorithmic Dynamics Lab, Department of Oncology-Pathology & Center of Molecular Medicine, Karolinska Institute, Stockholm, Sweden

Witold Pedrycz, PhD
Department of Electrical and Computer Engineering, University of Alberta, Canada

Volumes in Series

For more information about the UCTDI series, please visit: https://www.elsevier.com/books-and-journals/book-series/uncertainty-computational-techniques-and-decision-intelligence

REACHABLE SETS OF DYNAMIC SYSTEMS

Uncertainty, Sensitivity, and Complex Dynamics

STANISLAW RACZYNSKI
Panamerican University
Faculty of Engineering
Mexico City, Mexico

ACADEMIC PRESS
An imprint of Elsevier

ISBN: 978-0-443-13384-8

For information on all Academic Press publications
visit our website at https://www.elsevier.com/books-and-journals

Publisher: Mara E. Conner
Editorial Project Manager: Mason Malloy
Production Project Manager: Omer Mukthar
Cover Designer: Matthew Limbert

Typeset by VTeX

Working together
to grow libraries in
developing countries

www.elsevier.com • www.bookaid.org

Contents

About the author

Stanislaw Raczynski

Stanislaw Raczynski received his master degree (1964) from the Electrical Engineering Department of the Academy of Mining and Metallurgy (AGH) in Krakow, Poland, and his PhD (1969) and habilitation degrees (1977) from the same institute, in the area of control theory and optimization methods. In 1964 Dr. Raczynski joined the Institute for Automatics and Industrial Electronics of the AGH. From 1971 through 1972, he was head of the Computer Center of the AGH. Between 1973 and 1976 he worked as a researcher in the International Research Group in Moscow, USSR. In 1976 Dr. Raczynski became head of the Systems Analysis Group at the AGH. From 1980 through 1983 he participated in the activities of the European Workshop on Industrial Computer Systems. Between 1983 and 1986 he was a visiting professor of the National University of Mexico. In 1986 Dr. Raczynski joined Panamericana University in Mexico City. Between 1996 and 2000 and then between 2002 and 2004 Dr. Raczynski was the International Director of The Society for Computer Simulation. He wrote four books on computer simulation (Wiley, Springer Nature, UK, and LIMUSA, Mexico) and has more than 140 articles and papers published in professional journals and conference proceedings.

Preface

As in many fields of science and technology, the things that delay progress are mostly paradigms that can hardly be overcome and changed. One such paradigm is the strange conviction that everything that happens in the real world and is continuous in time can be described by differential equations. Moreover, many simulationists believe that industrial or urban dynamics can be described in the same way as electric circuits. In this book, we focus on *differential inclusion*, which is a generalization of a differential equation and may be used as a modeling tool when treating uncertainty.

I agree that systems dynamics, founded by Jay Forrester, and based on differential and difference equations, has been a great contribution to modeling and simulation and still provides valuable ideas to managers, leaders, and engineers. However, it should be revised, not only for being a 60-year-old tool.

Differential inclusions are closely related to differential equations. So, we remain in the field of continuous simulation in this book. Observe that in a digital computer nothing is continuous. Consequently, the continuous simulation of this hardware is an illusion. Despite this, there are, of course, many implementations that reflect the continuous reality with acceptable precision. As for precision, it is always limited by the hardware. The finite length of the binary representation of real numbers is one of the sources of uncertainty. This could be called micro-uncertainty, like the Heisenberg uncertainty principle in physics. Anyway, uncertainty is always present in the real world, as well as in computer simulation. The uncertainty caused by the finite real number resolution is very small, but it should not be forgotten. The errors caused by the computer resolution are reduced by the use of sophisticated numerical methods, but they are always present.

Finally, there are many other sources of uncertainty, including erroneous information generated intentionally. Uncertain parameters are frequently treated as random variables. However, not all that is uncertain is necessarily random. In this book, while treating uncertainty, we take into account also the *tychastic* variables, which have no expected value, variance, or other probabilistic properties. Such kind of uncertainty is explained in Chapter 1. The term *tychastic* comes from the works of Aubin and Saint-Pierre, referenced in Chapter 1.

Looking at the annals of the history of computer simulation, observe that the only machines that carried out truly continuous simulation were analog computers. These machines, commonly used in the middle of the 20th century, had a great impact on modeling and simulation in many fields of science and engineering. By the way, let me note that an analog electronic solver of differential inclusions has been built in the mid-1960s, based on operational amplifiers. This machine was a part of a PhD project at the Academy of Mining and Metallurgy in Krakow.

Microuncertainty is not the main problem in modeling and simulation. Sometimes, the simulationist reduces his/her conceptual horizon to an idealized continuous and deterministic version of the model. Then, comparing the model to the real world and looking for the model parameters, he/she faces the problem of uncertainty in the real world. There are mainly uncertain parameters, and sometimes the model structure is also uncertain. Another type of uncertainty is related to predictions that may influence the decision variables and controls.

Uncertain model parameters are frequently treated as random variables and incorporated into a stochastic version of the model. This approach provides tools that tell us how the probability of a certain model outcome can change over time. This may represent important information about possible changes in the real world. However, this "stochastic" paradigm may fail. *An uncertain parameter is not the same as a random-valued parameter.* To use probabilistic methods, we must know something about the probabilistic characteristics of the simulated object. The most required properties are the expected value, variance, or probability distribution. There are situations where such data are hardly available, or they do not exist at all. For example, information on the price of a share on the stock market is frequently uncertain, but it is not a random variable. This may be false information published intentionally to confuse the stock market runners. The such parameter has no expected value and no probability distribution. When implementing a simulation tool, a more realistic question concerns the possible limits for the values of the parameter. Such simulations may provide a rough assessment of the real system behavior, but in many cases, this may be the only way to obtain reasonable results. This way of treating uncertainty leads to applications of differential inclusions. This approach to uncertainty treatment is also referred to as the *tychastic* approach, mentioned before. See Aubin et al. (2014) and Aubin and Saint-Pierre (2005) in the references of Chapter 1.

Another concept discussed here is the *functional sensitivity*. While the conventional sensitivity is defined in terms of partial derivatives of model output with respect to the parameters, we define the *local functional sensitivity* using concepts of the calculus of variations. It is pointed out that non-local functional sensitivity analysis is, in fact, the search for the reachable sets of a differential inclusion related to the model.

The definitions, assumptions, and some mathematical concepts related to differential inclusions are explained in Chapter 1. The differential inclusion is a known tool, but it is not very popular among those who work in computer simulation and model building. The differential inclusion is a generalization of a differential equation. An ordinary differential equation in its state space or canonical form has a left-hand side that is a (scalar or vector) state space derivative and a right-hand side that is a function of the state variables, parameters, external perturbations, and time. The right-hand side of a differential inclusion is a *set-valued function*. If the right-hand side can be parametrized by some auxiliary variables, it can be expressed in the form of a corresponding control system. Roughly speaking, a function whose derivative is included in the DI's right-hand side over a given time interval is a trajectory of the DI. The trajectories are commonly treated as solutions to the DI. However, in this book we use the term *reachable set* to refer to the solution. The reachable or *attainable* set is the union of the graphs of all trajectories of the DI. To avoid confusion or conflict of terminology, we will not use the term solution, and rather talk about trajectories and reachable sets.

Chapter 2 describes the main tool we use in other chapters, called the *differential inclusion solver*. This is an algorithm and software developed by the author, which calculates and displays the images of reachable sets. A commonly made error in reachable set computing is to explore the interior of the set to assess its shape. The differential inclusion solver generates a series of trajectories that scan the *boundary*, and not the *interior*, of the reachable set. The same algorithm determines the reachable sets of dynamic models defined in terms of control systems.

DIs are closely related to optimal control theory. For this reason, Chapter 3 is dedicated to the concepts and applications of optimal control methods.

Other chapters describe applications of the solver in diverse fields of engineering, marketing, and other models. The reachable sets for the models are shown for fluctuating perturbations and are used in the sensitivity analysis. Chapter 4 is dedicated to stock market dynamics. In Chapter 5 reachable sets for flight maneuvers are shown. A similar problem for ship control and maneuvers is discussed in Chapter 6.

In Chapter 7 we obtain the reachable sets for some mechanical systems. The dynamics and the reachable sets of a control circuit with a PID controller are analyzed in Chapter 8. The real PID is used with saturation and anti-windup. Reachable sets are calculated, which may be used in sensitivity analysis and robust control design.

Another example of a control system with a strongly non-linear controlled object (induction motor) is discussed in Chapter 9. The reachable sets for V/f speed control of an induction motor are shown.

Chapter 10 is dedicated to models of the spread of epidemics. Finally, in Chapter 11 we consider the somewhat abstract problem of the "trip to the future" and ideal predictors. The uncertainty of the information taken from the future is treated as a permissible set that allows to define a differential inclusion for the problem. It is pointed out that in this way we can solve the problem of the ideal predictor.

This diversity also reflects the interdisciplinary nature of the research field named modeling and simulation. To be a modeler and simulationist, one must not only dominate the use of simulation tools, but also know how a plane flies, how shoes are produced, how an epidemic disease spreads, and a lot of other things.

Development and validation of the models used here are not topics of this book. Most of the models are taken from the literature, together with their parameters. What is perhaps new is the *treatment of the uncertainty of parameters*, carried out by means of the differential inclusions, and additional dynamics and inertia inserted into the model. The main result is the reachable set, instead of a single trajectory provided by simple simulations. Such type of results may provide a wider view of complex model dynamics.

For the reader's convenience, a short section on differential inclusions is included in several chapters. Thus, each chapter can be read as an independent unit. However, remember that the book is not a collection of articles. It is an integrated work, where the chapters reference to each other and present a unified approach to the discussed models.

Acknowledgments

In this book, updated versions or fragments of my previously published articles have been used, as listed below.

Simulation and optimization in marketing: Optimal control of consumer goodwill, price and investment. *International Journal of Modeling, Simulation, and Scientific Computing*, World Scientific Publishing Company, 2014; 5 (3).

A Market Model: Uncertainty and Reachable Sets. *International Journal for Simulation and Multidisciplinary Design Optimization*, EDP Sciences, 2015; 6 (A2), ISBN/ISSN: (Electronic Edition): 1779-6288.

Differential inclusion approach to the stock market dynamics and uncertainty. *Economy, Computing, Optimization, Risks, Finance, Administration, and Net Business*, 2015; 1 (1), 58–65, ISBN/ISSN: 2444-3204.

CHAPTER ONE

Differential inclusions

Abstract

In this chapter you will find some information on the history of differential inclusions and remarks related to terminology. Readers not interested in the mathematical details of the concepts used in further chapters may skip this chapter. Anyway, in each of the next chapters a short section on differential inclusions is included. The basic definitions are recalled and the properties of differential inclusions and the corresponding reachable sets are discussed. The connection between differential inclusions and control systems is pointed out.

An interpretation in the language of the theory of categories is given. Reversibility and the inverse problem are discussed. We point out that models with uncertainty are not reversible (they cannot be solved by simply reversing the time in the equations).

The discrete event version of the differential inclusion, named *discrete decision inclusion*, is mentioned, and some corresponding simulation results are presented.

Keywords

Differential inclusion, Uncertainty, Reachable set

1.1. History and terminology

Let us recall some early works on differential inclusions (DIs). More detailed assumptions will be given further on in this chapter. A *DI* is defined by the following statement:

$$\frac{dx}{dt} \in F(x, t), \ x(0) \in X_0, \tag{1.1}$$

where t is a real variable (representing time in this book), x is a function of time, $x(t) \in R^n$, F is a mapping from $R^n \times R$ to subsets of R^n, and $X_0 \subset R^n$ is the initial set. R^n is the real n-dimensional Euclidean space. In the following, R is equal to R^1. F is also called the *set of admissible directions*. Note that here we limit our discussion to the real Euclidean space. In more advanced research on DIs, the state variable may be considered as an element of more abstract spaces.

Let us recall some historical facts.

Early works on DIs in the finite-dimensional case were published by Marchaud [25] and Zaremba [48]. The terminology used in those papers was slightly different from that used in more recent works. Zaremba uses the term *paratingent equation* and Marchaud calls the same *contingent equation*. Both of them did not suspect the importance of these concepts for future applications in control theory. Moreover, the research was criticized by contemporary mathematicians and physicists as abstract and useless work. In fact, Zaremba and Marchaud had already described the main properties of trajectories and reachable sets of control systems with convex sets of admissible directions many years before those problems were treated by control theory. Further works of Turowicz and Wazewski provided more important results. Wazewski [46,44,45,43], Turowicz [42,41], and Plis [30] gave the generalization of the basic results to the case of non-convex sets of admissible directions (the set F in (1.1)). This is frequently the case in control systems, in particular when *"bang-bang"* type control is used. In the terminology used by Wazewski, the right-hand side of the DI is called *orientor field* and (1.1) is called the *orientor condition*.

One of the most important concepts, widely discussed by Wazewski, is the *quasitrajectory* of an orientor field, defined in the next section (a weak solution to the DI). This concept is closely related to "bang-bang" control and *sliding regimes* known in automatic control theory. Wazewski [45] discussed the DIs derived from a control system, where the control variable belongs to a set $C(t) \subset R^m$. In this case, the set F on the right-hand side of the corresponding DI can be obtained by parametrization, F being a mapping of C. In the article of Wazewski, both C and F need not be convex. This relation to control systems will be discussed in more detail further on. Some important work on optimal and non-convex control done in the early 1960s was inspired by the activities of the Department of Automatic Control of professor Henryk Gorecki, Academy of Mining and Metallurgy in Krakow, with close collaboration with the Krakow Division of the Institute of Mathematics of the Polish Academy of Sciences.

The concept of the quasitrajectory in the case of the non-convex right-hand side of (1.1) has been used later (Sentis [36] and Krbec [23]). See also the book on DIs by Aubin and Cellina [4]. For remarks on optimal control in abstract spaces, consult [34,33].

Some relevant research on DIs in *game theory* has been done. Many of the works in the field use the Hamilton–Jacobi–Bellman equations and the methods of control theory closely related to DIs. See for example Tsunumi and Mino [40], who worked on the Markov perfect equilibrium problems in differential games. Grigorieva and Ushakov [19] consider the differential game of pursuit–evasion over a fixed time segment. A more general, variational approach to differential games can be found in [5]. DIs were used by

Solan and Vieille [38] to study equilibrium payoffs in quitting games. For general problems of game theory, consult Petrosjan and Zenkevich [29], Isaacs [22], or Fudenberg and Tirole [17].

1.2. Some definitions

Before discussing solutions to DIs and their applications, let us recall some elemental concepts from mathematical analysis. We limit the definitions to the real Euclidean n-dimensional space R^n.

In Wazewski's original works on DIs, the field of directions in R^n, defined by the multivalued function $F(x, t)$ in (1.1), is called the *orientor field*. As Wazewski's terminology is not commonly used in recent publications, we will call this field simply F when no relevant ambiguity arises (do not confuse F with the field concept of advanced algebra).

In the following, $I = [a, b]$ is a non-empty interval of R. The abbreviation "a.e." means "almost everywhere," i.e., everywhere in I, except a set of measure zero. A set of measure zero defined over an interval of R is a set which can be covered by a finite or enumerable sequence of open intervals whose total length (i.e., the sum of the individual lengths) is arbitrarily small. A function $y : R^n \to R^n$ satisfies the *Lipschitz condition* if a real constant L exists such that for all $z_1, z_2 \in R^n$

$$|y(z_1) - y(z_2)| < L|z_1 - z_2|, \tag{1.2}$$

where the real constant L is independent of z_1 and z_2. The Lipschitz condition plays important role in the problems of existence and uniqueness of the solutions to differential equations.

A function $y : R^n \times I \to R^n$ satisfies the *generalized Lipschitz condition* if

$$|y(z_1, t) - y(z_2, t)| < K(t)|z_1 - z_2|, \tag{1.3}$$

where $t \in I$ and $K(t)$ is a real, non-negative, and Lebesgue integrable function on I. This condition has been used in some early works on DIs, mainly in the problems of existence, continuity, and other properties of reachable sets.

A function $y : I \to R^n$ is called *absolutely continuous* if for every $\varepsilon > 0$ a number δ exists such that for any finite collection of disjoint subintervals $[\alpha_k, \beta_k]$ of I satisfying the inequality

$$\sum (\beta_k - \alpha_k) < \delta, \tag{1.4}$$

we have

$$\sum |\gamma(\beta_k) - \gamma(\alpha_k)| < \varepsilon. \tag{1.5}$$

An absolutely continuous function is continuous.

The point-to-set distance between a point $x \in R^n$ and a set $A \subset R^n$ is defined as

$$d(x, A) = d(A, x) = \inf\{|x - b| : b \in A\}, \tag{1.6}$$

where inf means the greatest lower bound.

Let X and Y be two non-empty subsets of a metric space. The Hausdorff distance between X and Y is defined as follows:

$$d(X, Y) = max\left[\sup_{x \in X} d(x, Y), \sup_{y \in Y} d(X, y) \right], \tag{1.7}$$

where sup represents the least upper bound and $d(*, *)$ is the point–to–set distance.

The Hausdorff distance permits to use the concept of continuity for set-valued functions. We say that a mapping from the real numbers to the space of closed subsets of R^n is *continuous in the Hausdorff sense* if it is continuous in the sense of the Hausdorff distance (in the topology induced by the Hausdorff distance).

Suppose that f is a real-valued function $f : R^n \to R$, $x_o \in R^n$, and $f(x_o)$ has a finite value in x_o. The function f is *upper semi-continuous* at x_o if for every $\varepsilon > 0$ there exists a neighborhood U of x_o such that

$$f(x) \leq f(x_o) + \varepsilon \ \forall \ x \in U. \tag{1.8}$$

Here, we limit ourselves to finite values of x. In a similar way, the function f is *lower semi-continuous* at x_o if for every $\varepsilon > 0$ there exists a neighborhood U of x_o such that

$$f(x) \geq f(x_o) - \varepsilon \ \forall \ x \in U. \tag{1.9}$$

A function is upper or lower semi-continuous if the above conditions hold for all x_o in the interval under consideration, respectively. Fig. 1.1 shows an example of a lower semi-continuous function. Note that the function value at x_o is defined as shown by the black dot. Roughly speaking, the function cannot "jump" to lower values in any point where it is defined.

Continuity can also be defined for set-valued functions, using the metric induced by the Hausdorff set-to-set distance. As for lower semi-continuity,

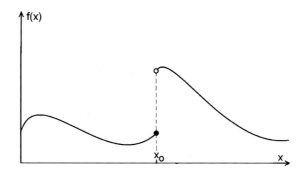

Figure 1.1 A lower semi-continuous function.

the following definition is used. Let F be a mapping from R^n to subsets of R^n. We say that F is lower semi-continuous at $x_o \in R^n$ if and only if for any open set $V \subset R^n$ such that $F(x_o) \cup V \neq \emptyset$, there exists a neighborhood $U \subset R^n$ of x_o such that

$$\forall x \in U : F(x) \cup V \neq \emptyset. \tag{1.10}$$

The mapping F is said to be lower semi-continuous on R^n if F is lower semi-continuous for every $x \in R^n$. Lower semi-continuity is an important property of certain set-valued functions associated with the original function F of (1.1).

Consider a mapping $F : R^n \times I \to P(R^n)$, where $P(X)$ denotes the power set, i.e., the set of all subsets of a space $X, I \subset R$. A selection (or selector) of F over I is a function $z(t)$ such that $z(t) \in F(x, t) \ \forall t \in I$. The existence of selectors is a consequence of the known axiom of choice (Halmos [20]).

Some facts related to selections are quite interesting and may contradict our intuition. For example, it is not true that a continuous field should have a continuous selector. Aubin and Cellina [4] show an example of such field, as follows (the example is slightly modified, exact formulae are omitted).

Consider a mapping $F : [0, 2] \to P(R^2)$, which is an ellipse with a "hole" in it, as shown in Fig. 1.2. Both the small axis (x_1-direction) of the ellipse and the size of the hole tend to zero as t approaches one, and the hole rotates with angular speed equal to $1/|t - 1|$. The hole and the small axis disappear and F reduces to a vertical section at $t = 1$. It is easy to see that the field F is continuous in the Hausdorff metrics. Now, consider a selector of F and suppose that it is continuous. As the selector is supposed to be continuous, it cannot "jump" from one side of the hole to another.

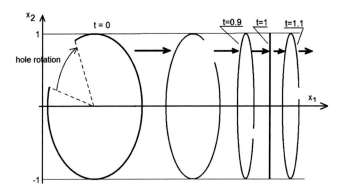

Figure 1.2 Continuous field F that has no continuous selection.

The movement of the hole forces the component x_2 to oscillate between -1 and 1, with a frequency that tends to infinity while t approaches one. This means that the selector has oscillating discontinuity at $t = 1$.

On the other hand, a discontinuous field may have continuous selections. For example, let the value of $F : [0, 2] \to R^2$ be a filled rectangle with vertices $(-1, 1)$, $(1, 1)$, $(1, -1)$, and $(-1, -1)$ for $t < 1$ and a filled rectangle with vertices $(-0.5, 0.5)$, $(0.5, 0.5)$, $(0.5, -0.5)$, and $(-0.5, 0.5)$ for $t > 1$.

The field is discontinuous at $t = 1$. However, it has a continuous selection $z(t) \equiv 0$.

Let U be a compact separable metric space (in our case a closed and bounded subset of R^n). Assume that there exist an interval I and an absolutely continuous function $x(t) : I \to R^n$ such that a.e. in I, $x'(t) \in \{f(x(t), u) : u \in U\}$ (x' stands for dx/dt). Then there exists a measurable $u : I \to U$ such that

$$x'(t) = f(x(t), u(t)) \text{ a.e. in } I \tag{1.11}$$

(Aubin and Cellina [4, Chapter 1, Section 14, Corollary 1]). This corollary may be useful when dealing with control systems.

Consider a selector $z(t)$ of a DI and a function $x(t)$ such that

$$x'(t) = z(t) \text{ a.e. in } [t_0, t_1], \ t_0 < t_1. \tag{1.12}$$

The solution to this equation is a function $x(t)$ that is measurable and differentiable a.e. and fulfills the equation a.e. over a given time interval I. We will call the function $x(t)$ a trajectory of the DI.

1.2.1 The reachable set

Roughly speaking, the *reachable* (or *attainable*) *set* of a DI is the union of the graphs of all trajectories of the DI. A more exact definition is as follows.

First, recall that the *graph* of a function $f(t)$ is the set of all ordered pairs $(t, f(t))$.

Let X_0 be a closed and connected subset of R^n, let $I \subset R$ denote an interval $[t_0, t_1]$, let $x(t) \in R^n$ be the model state vector, and let $F : R^n \times R \to P(R^n)$ be a set-valued function, where $P(X)$ is the power set, i.e., the set of all subsets of a space X.

The *reachable* or *attainable set* of (1.1) is defined as the union of the graphs of all trajectories of (1.1).

The term *emission zone* has been used in the early works. Here, we will rather use the term *reachable* or *attainable set*. In many works on DIs, the mapping F is called a field of permissible directions and a trajectory of the DI is also called a trajectory of the field F.

Let us comment on the term "solution to the DI." It is commonly understood that a trajectory of the DI is its solution. Observe that a DI normally has an infinite number of trajectories. Thus, the trajectory cannot be just called "the solution." Our point is that *the* solution to a DI is given by its reachable set. Consider a sequence of DIs with shrinking right-hand side that, in the limit, degenerates to a single-valued function. Then, the corresponding sequence of reachable sets tends to the graph of the solution of the resulting differential equation. This is an argument to call the reachable set the solution to the DI. However, to avoid ambiguity and conflict of terms, the term "solution to a DI" will not be used in the following sections. Instead, we will discuss trajectories and reachable or attainable sets.

An absolutely continuous function $x(t)$ is called a *quasitrajectory* of the DI (1.1) over an interval I with initial condition x_0 if a sequence of absolutely continuous functions $\{x_i\}$ exists such that

$$\begin{cases} (i) & x_i(t) \to x(t) \ \forall \ t \in I = [t_0, t_1], \\ (ii) & d(x'(t), F(x_i(t), t)) \to 0, \text{ a.e. on } I, \\ (iii) & x_i' \text{ are equibounded on } I, \\ (iv) & x(0) = x_0. \end{cases} \tag{1.13}$$

Recall that a sequence of functions $\{x_i(t)\}$, $t_o < t_1$, is equibounded if such M exists that

$$|x_k(t) \leq M \ \forall \ k = 1, 2, 3 \ldots \ \wedge \ t \in [t_0, t_1]. \tag{1.14}$$

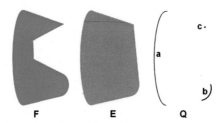

Figure 1.3 Example of the sets *F*, *E*, and *Q*.

A function $x = x(t)$ is called a *strong quasitrajectory* of the field F if there exists a sequence $x_i(t)$ $(i = 1, 2, ...)$ of trajectories of F such that $x_i(t) \rightarrow x(t)$.

1.2.2 The tendor set

Turowicz [42] has given sufficient conditions for a quasitrajectory to be a strong quasitrajectory. The notion of strong quasitrajectory ("sliding regime") was introduced independently by Filippov [12] under stronger hypotheses. The set $E = conv(F)$ is defined as the smallest convex and closed hull of the set F. The set $Q = tend(F)$ is the smallest closet subset of F that has the same convex hull, $conv(Q) = conv(F)$. Fig. 1.3 shows an example. Note that the tendor set Q contains the curved lines a and b and the point c.

A very useful property of quasitrajectories of the fields F, E, and Q was found by Wazewski. He pointed out that if the field F is continuous in the Hausdorff sense, then the field E is also continuous and the field Q is lower semi-continuous. The most important property of these fields is that under some additional regularity conditions, the fields F, E, and Q have the same quasitrajectories. Moreover, the Filippov–Wazewski theorem states that if F satisfies the Lipschitz condition, then for each quasitrajectory a sequence of trajectories exists that converges to that trajectory. Consequently, the three fields have the same reachable sets with accuracy to their closures (reachable sets of the fields F and Q need not be closed). In other words, the sets of trajectories of the inclusions

$$x'(t) \in F(x(t), t) \text{ and } x'(t) \in Q(x(t), t) \tag{1.15}$$

are dense in the set of trajectories of the inclusion

$$x'(t) \in E(x(t), t). \tag{1.16}$$

The word "dense" means that in any neighborhood of a trajectory of the first inclusion there exists a trajectory of the second. This also means that in many cases of control systems, the "tendor" or "bang-bang" type of control (field Q) can be used without restricting the system's reachable set. See also [34,33]. Let us denote the set of all trajectories of the field F as $\{F\}$ and the set of all its quasitrajectories as $\{F\}^*$. By $comp(R^n)$ we denote the collection of all non-empty compact subsets of R^n.

Let $T = [-\infty, \infty]$, $W = R^n \times T$. Consider the following hypothesis.

Hypothesis H(F). For each $(x, t) \in W$, $F(x, t) \in comp(R^n)$, $F(x, t)$ is bounded and continuous on W.

Wazewski [43] pointed out that under Hypothesis H(F) we have

$$\{F\}^* = \{Q\}^* = \{E\}^* = \{E\}. \tag{1.17}$$

Let $W = R^n \times T$, $E \subset W, \Theta(k)$ be the hyperplane $t = k$, and let $S(E, k) = \{E\} \cap \Theta(k)$ (the "time section" of $\{E\}$). One of the results shown by Zaremba (1936) is that $S(E, k)$ is a compact and connected set. Thus, the same property holds for the sets of quasitrajectories of the fields F and Q. A more detailed discussion on this and other properties of reachable sets can be found in Wazewski [45]. For the field E it was also shown that if a point in $I \times R^n$ is accessible from the initial point $x(0)$ of (1.1), then it is also accessible in optimal (minimal) time.

1.3. Differential inclusions and control systems

DIs are closely related to control systems. To see this, consider the dynamical system

$$\begin{cases} \dfrac{dx}{dt} = f(x, u, t), \\ x(0) = x_0, \ u(t) \in C(x(t), t), \\ x \in R^n, \ u \in R^m, \ C(x(t), t) \subset R^m, \end{cases} \tag{1.18}$$

where $x(t)$ is the n-dimensional system state, $u(t)$ is the m-dimensional control variable, and the set C represents the control restrictions. Define the function F as follows:

$$F(x, t) = \{z : z = f(x, u, t) : u \in C(x, t)\}. \tag{1.19}$$

In other words, the function $f(x, *, t)$ maps $C(x, t)$ into the set $F(x, t)$ for each fixed x and t. Using this set-valued function F in the DI (1.1), we obtain the DI derived from the control system (1.18):

$$\frac{dx}{dt} \in F(x(t), t), \; x(0) = x_0. \tag{1.20}$$

In (1.20), the control variable does not appear explicitly. Control system (1.18) and the differential inclusion (1.20) have the same trajectories.

Define the *bang-bang kernel* of $C(x, t)$ as follows:

$$B(x, t) = \{u : u \in C(x, t), f(x, u, t) \in Q(x, t)\}, \tag{1.21}$$

where Q is the corresponding tendor set of F.

The consequence of equality (1.17) is that we can use the bang-bang kernel B instead of C to obtain a control system with the same quasitrajectories. This means that we can hit or approximate any point inside the reachable set of (1.20), using the restricted control set B that contains fewer points than C. In many practical applications B reduces to a finite number of points. This permits us to simply use bang-bang control instead of continuous control, with less complicated instrumentation.

As a consequence of the above remarks, we may conclude that given a DI, we can find the corresponding control system by parametrizing the function F with a certain parameter u (control variable). This is true in many cases. However, the parametrization problem is not so simple. Consult Aubin and Cellina [4, Chapter 1, Section 7, "Application: The parametrization problem"]. In that section, it is pointed out that the existence of continuous selection of F is not sufficient to enable parametrization of F. Fortunately, the mappings considered in the following chapters of this book are regular enough to permit parametrization.

1.4. Uncertainty, differential games, and optimal control

As stated before, our point is that *not* everything that happens in the real world and is continuous in time can be described by differential equations. Thus, the differential equation approach to modeling is somewhat dangerous because it forces the modeler to look for something (a differential equation model) that might not exist at all. Such efforts may result in attempts to change the real world to fit our simulation tools, while the correct way should be the opposite. The challenge is to look for new tools or to use those that have been known for a long time but are not used or simply forgotten.

There are many theoretical results in the field of DI-based models, but little has been done in the area of numerical methods and practical applications. The solution to a DI (the reachable set) needs hundreds of single system trajectory evaluations, which makes the whole task computationally expensive (see Chapter 2). Another challenge is the representation of the results (n-dimensional sets and boundary surfaces).

The absence of reliable data in computer simulations is a big obstacle in many simulation projects. Models that are nice and valid from the academic point of view often result to be useless in practical applications when the end user cannot provide reliable information to adjust the model parameters. A common way to deal with this lack of exact data is to suppose that some model parameters or input variables are random. This approach results in a stochastic model, where every realization of the system trajectory is different and the problem is to determine the probability density function in the system space for certain time sections, the variance, confidence intervals, etc.

Such stochastic analysis is interesting and useful, but not always possible. Some parameters of the model have uncertain values, and the model user may have no idea about their probabilistic properties. He/she is more likely to provide an interval for the values of an uncertain parameter than the corresponding probability distribution. Some external perturbations can fluctuate within certain intervals, and the task consists in obtaining the interval for some output variables. Frequently, the user wants to know a possible extreme value rather than the probability to reach it (recall the law of Murphy!). The uncertainty treatment in this book has nothing, or very little, to do with "Monte Carlo" or stochastic simulation. The intervals we are looking for are not confidence intervals or any other statistics.

The need for a new approach, different from the stochastic uncertainty treatment, has recently been commented on by some authors. The non-stochastic approach to uncertainty treatment is referred to as *tychastic* (Aubin et al. [2] and Aubin and Saint-Pierre [3]). In Aubin et al. [2], the authors observe that the *"tychastic viability measure of risk"* is an evolutionary alternative to statistical measures. As an example, the price intervals are used instead of probability distributions in the solvency capital requirement problem, dealing with evolution under uncertainty. Aubin and Saint-Pierre [3] use tychastic variables to obtain the corresponding properties for portfolios and evaluate the capital and the transaction rule.

Some problems of differential games can also be dealt with using DIs. A missile that follows a moving target will fail when the target trajectory escapes from the attainable set of the missile. This can be used to verify different strategies of both players involved in the game.

While modeling systems with uncertainties, an input signal or a disturbance is often an unknown function that belongs to a given restriction set. In this case, the system can be modeled by the corresponding DI. This could be an alternative and conceptually different approach to modeling of systems with uncertain parameters (deterministic instead of stochastic). If we admit the uncertain parameter to be a random variable, then the problem is to determine the resulting density functions in the state space over a given time interval. However, we cannot obtain any result without knowing the probabilistic properties of the input functions. In many real situations we have vague or erroneous information on the distributions and spectra of the input signals or perturbations. This can make the simulation impossible or lead to wrong results. As stated before, the uncertain parameter may represent erroneous information, intentionally inserted into the real system, which means that it cannot be treated as a random one. Other difficulties appear when generating multidimensional random variables with arbitrary density functions.

On the other hand, using a DI model, we only need to know the restrictions for the uncertain signals. As the result, we get the system attainable set, which can provide important information on possible system behavior. While simulating robust control systems, we often want the system to remain in some permissible region of states, regardless of possible disturbances. The reachable sets of the corresponding DIs can provide solutions to such problems.

As for the control systems, DIs are closely related to the optimal control problem. From the theory of optimal control, it is known that the optimal trajectories belong to the boundary of the system's reachable set; see Lee and Markus [24], Mordukhovich [26], and Pontryagin [31]. The algorithm of optimal control and its applications will be discussed in Chapter 3.

Another application is a new approach to system sensitivity, named *functional sensitivity*. This concept is discussed in Section 1.5.

In this book, while introducing uncertainty, we assume that some model parameters become functions of time. However, we should remember that in some cases, this is not exactly true. If we use a model that has been developed assuming constant parameters, it is not always possible to simply replace a parameter with a function of time. To do this, we should check if the derivative of the parameter explicitly intervenes in the model or not. However, in the models discussed here, we do not take such problems into account, as this would require repetition of the task of model building. We do not do this because the topic of this book is the methodology of determination of reachable sets, and not model development.

1.5. Functional sensitivity

Classical sensitivity analysis (SA) (the basic local version) is based on the partial derivatives of the model output Y, with respect to components of an input vector (model parameters) $u = (u_1, u_2, ..., u_n)$, at a given point u_0:

$$\left| \frac{\partial Y}{\partial u} \right|_{u_0}. \tag{1.22}$$

The derivative is taken at some fixed point in the space of the input (hence the "local" in the name of the analysis mode). The use of partial derivatives suggests that we consider small perturbations of the input vector, around the point of interest u_0. Consult Cacuci [6]. The classical approach to SA and its modifications are not the topic of the present book, so here we will not give any extensive overview, but make only some remarks.

Scatter plots represent a useful tool in SA. Plots of the output variable against individual input variables after (randomly) sampling the model over its input distributions are made. This gives us a graphical view of the model sensitivity. Consult Friendly and Denis [16] for more details.

Regression analysis is a powerful tool for sensitivity problems. It allows us to examine the relationship between two or more variables of interest. The method is used to analyze the relationship between a response variable and one or more predictor variables or perturbations. See, for example, Freedman [15] or Cook [8].

For non-linear models, one of the useful tools is the *variance-based* or *Sobol* method. It decomposes the variance of the output of the system model into fractions which can be attributed to the input or sets of inputs. Consult Sobol [37].

The *screening method* is used to identify which input variables are contributing significantly to the model output in high-dimensionality models, rather than exactly quantifying sensitivity (i.e., in terms of variance). The method is relatively simple and useful in preliminary analysis. Consult Campolongo and Saltelli [7] and Morris [27].

The *logarithmic gain* is a normalized sensitivity measure defined by the percentage response of a dependent variable to an infinitesimal percentage change in an independent variable. In dynamical systems, the logarithmic gain can vary with time, and this time-varying sensitivity is called dynamic logarithmic gain. This concept is used in dynamic SA, where the core model is a dynamic system, described by ordinary differential equations (ODEs). Consult Sriyudthsak et al. [39].

System dynamics software offers tools for SA of dynamic models. Programs like Vensim or PowerSim include procedures that generate multiple model trajectories where the selected model parameters vary from one trajectory to another. However, in these packages the parameters are constant along the trajectory. Our approach is different. As explained in the following sections, we treat the perturbations as functions of time. The main tool used here is the DI.

1.5.1 Functional sensitivity concept

Consider a dynamic model described by an ODE

$$\frac{dx}{dt} = f(x, u, t), \tag{1.23}$$

where $x = (x_1, x_2, ..., x_n)$ is the state vector, $u = (u_1, u_2, ..., u_m)$ is the perturbation (parameters, control) vector, and t is the time. We have $x \in X$, $u \in U$, $f : X \times U \times R \to X$. Here, X is the state space, U is the control space, and R is the real number space. We restrict the considerations to the case $X = R^n$, $U = R^m$, $R = R^1$, R^k being the real Euclidean spaces. Let $t \in I = [0, T]$ and let G be the space of all measurable functions $u : I \to R^m$.

Now, consider a variation δu of u and a perturbed control u^*. The variation is a function of time, so that $u^*(t) = u(t) + \delta u(t) \; \forall t \in I$. The solution to (1.23) over I with given initial condition $x = x_0$ and given function $u(t)$ will be called a trajectory of (1.23). Thus, any component x_k of the final value of $x(t)$ depends on the shape of the whole function u. In other words, $x_k(t) = x_k(t)[u^*]$ (for any fixed t) is a *functional* (not a function) of u^*. Unlike a function, in our case, a functional is a mapping from the space G to R. We denote $\delta x_k = x_k[u + \delta u] - x_k[u] = x_k[u^*] - x_k[u]$. In this book, the *local functional sensitivity* is defined as

$$S_k = \left| \frac{\partial x_k}{\partial u_0} \right|. \tag{1.24}$$

Note the difference between the conventional local sensitivity (1.22) and the functional sensitivity (1.24).

The notation δu is nothing new; it denotes the variation of the function u, as defined in the calculus of variations (Nearing [28] and Elsgolc [11]). The term (1.24) defines a local property of the trajectory $x_k(t)$. Here, we are rather interested in the response of the model to perturbations that are not necessarily small. We will not enter in the methodology of the variational calculus. Our task is to define the functional sensitivity as the set of the graphs of all trajectories of (1.23), where $u = u_0 + \Delta u$. Here, $\Delta u(t)$ is a limited perturbation. This is equivalent to saying that $u(t)$ belongs to a set of restrictions $C(t)$, $u(t) \in C(t)$, $\forall\, t \in I$. Here, $C(t)$ is a subset of R^m. When u scans all possible values inside the set C, the right-hand side of (1.23) defines a set-valued function. This way, (1.23) with disturbed control also defines the corresponding DI.

The functional sensitivity defined this way is non-local. We do not use the term "global," because this is not a global property of the model. We only do not require the perturbation to be small.

1.5.2 Example: a non-linear model

Consider a simple second-order non-linear model:

$$\begin{cases} \dfrac{dx_1}{dt} = x_2, \\ \dfrac{dx_2}{dt} = 1 - u_1 - x_1 - 0.1u_2(x_2 + 2.3x_2^2), \end{cases} \tag{1.25}$$

where u_1 and u_2 are uncertain parameters. Let the parameter u_1 fluctuate between -0.2 and $+0.2$ and let the parameter u_2 fluctuate between 0.025 and 0.175. The initial conditions are $x_1 = x_2 = 0$, and the final simulation time is equal to 10. Fig. 1.4 shows the 3D image of the reachable set. The three axes of the plot represent x_1, x_2, and the time. The image was generated by the DI solver described in Chapter 2. It can be seen that even for relatively small perturbations u_1 and u_2, the deviation of the state vector may be quite big.

Fig. 1.5A shows a comparison of the functional sensitivity to the conventional sensitivity provided in some system dynamics packages. The contour indicates the boundary of the reachable set for the model (1.25) at $t = 10$. These are end points of about 2000 boundary scanning trajectories. A small black region marked with an X is the result of the "Vensim-like" SA, where the parameters are constant along each trajectory. Region X

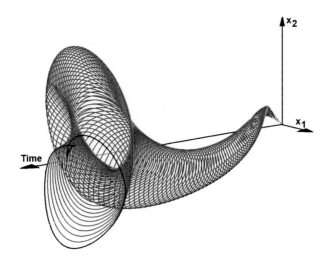

Figure 1.4 Functional sensitivity of model (1.25). The reachable set.

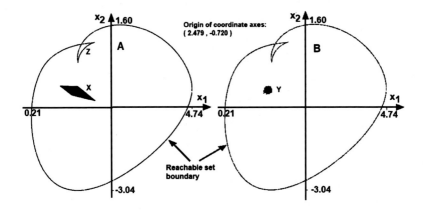

Figure 1.5 Final contour of the reachable set. Comparison with simple shooting.

was obtained by generating 50,000 trajectories with the same limits of uncertain parameters. In Fig. 1.5B we see region Y, obtained by generating
50,000 trajectories where the parameters could change their values randomly within the same limits at each integration step.

Note a small part of the contour marked with Z. These are points generated by the solver that belong to the interior of the reachable set. Recall
that the maximum principle provides the *necessary but not sufficient conditions*
for the trajectory to be optimal. Consequently, when the reachable set is

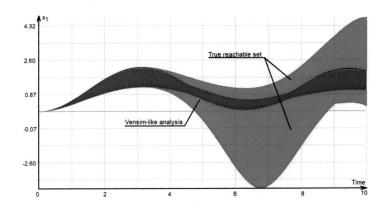

Figure 1.6 Functional sensitivity set projected into the x_1-time plane.

more complicated and folds several times, we can obtain also such extra points.

Fig. 1.6 shows a side view of the same reachable set, projected into the x_1-*time* plane. Note that the functional analysis region coincides with the conventional results for the initial time interval $[0, 35]$. However, the conventional analysis provides results very different from the true reachable set for greater time intervals.

1.6. Dualism and reachable sets

In this section, we generalize the model specification by using the language of the theory of categories (Fokkinga [14,13]). The main purpose of this theory is abstraction and generalization of the concepts being considered in different fields of mathematics. In our case of modeling and simulation, consider for example a model described by a DI. The categorical language makes no use of the continuous nature of the model. Thus, it can describe the uncertainty and the reachable sets of any other model, even of a discrete event type. This generalization should be exploited in more detail and with more possible applications in the future.

Perhaps the reverse problem discussed here does not require the use of category theory. However, we mention it here because it is an important generalization of concepts, not necessarily in mathematics. The suggestion that the task of modeling can be interpreted as a *functor* in the theory of categories may provide wider insight into the model building and validity issues.

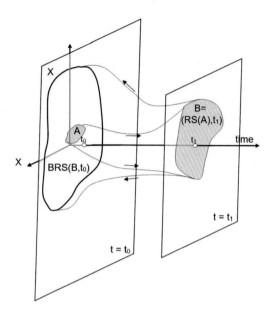

Figure 1.7 Initial set and the reachable set.

Recall that the reachable set of a DI (Fig. 1.7) with a given initial condition is the union of graphs of all trajectories going out of all the points of
the initial set $X(t_0)$ (with initial conditions belonging to $X(t_0)$). Note that
in this book, the trajectories are not considered solutions to the DI; they are
just trajectories. Let $I \subset R$, $I = [t_0, t_1]$. Thus, for a given DI, its reachable
set depends only on the initial set A and will be denoted as $RS(A)$. The
time section of the reachable set will be denoted in a similar way, with an
extra parameter representing the time instant. The term $RS(A, t_x)$ denotes
the intersection of $RS(A)$ with the plane $t = t_x$. Clearly, $A = RS(A, t_0)$.

Example. Consider a second-order system described by the following
equations:

$$\begin{cases} \dfrac{dx_1}{dt} = x_2, \\[2mm] \dfrac{dx_2}{dt} = u_1 - u_2 - x_1. \end{cases} \qquad (1.26)$$

The right-hand side of the second equation represents a set, parametrized
by two parameters u_1 and u_2, fluctuating by $\pm 10\%$. Fig. 1.8 shows the 3D
image of the reachable set in coordinates x_1, x_2, and t.

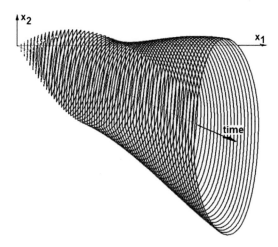

Figure 1.8 A 3D image of the reachable set of the differential inclusion derived from (1.26).

The reachable set was generated by the DI solver described in Chapter 2. The solver algorithm is somewhat complicated. It calculates the trajectories that scan the boundary and *not the interior of the reachable set*, using some concepts of optimal control theory. One could suppose that the reachable set can be obtained by a "random shooting," generating trajectories that correspond to selectors generated randomly. However, this is not true. Such random shooting can provide very confusing results, detecting only a small region inside the reachable set (see Chapter 2 for more details).

Eilenberg [10] and Mac Lane [35] introduced *categories*, as part of their work in algebraic topology, dated from the early 1940s. See also Asperti and Longo [1], Berkovitz [5], and Dutton [9]. As the concepts of the theory of categories provide a high level of abstraction, the language of the theory may be useful while constructing models and examine their properties. To consult other works that can be found in this and similar fields of model theory and category theory, see Fokkinga [14,13], Hoare [21], and Fudenberg and Tirole [17]. An interesting application in brain modeling can be found in Gomez and Sanz [18].

In this section we discuss the time-inverse problem in models with uncertainty. The model is given in the form of a DI, but as mentioned in the sequel, its properties in terms of categories are valid for any other model (may also be a discrete event) where the reachable set is defined. Following Fokkinga [14], we recall some definitions.

A *category* is defined by its data and the axioms the data must satisfy. The data are as follows:

1. A collection of things called *objects*. By default, $A, B, C, \ldots$ vary over objects.
2. A collection of things called *morphisms*, also called arrows. By default, $f, g, h, \ldots$ vary over morphisms.
3. A relation on morphisms and pairs of objects, called *typing* of the morphisms. By default, the relation is denoted $f : A \rightarrow B$ for morphism f and objects A, B. In this case, we also say that $A \rightarrow B$ is the *type* of f and that f is a morphism from A to B. Object A is called the *source* of f, and object B is the *target*. In other words, $srcf = A$ and $tgtf = B$.
4. A binary partial operation on morphisms, called composition. By default, $f; g$ is the notation of the composition of morphisms f and g.
5. For each object A there is a morphism called *identity* on A, idA. Note that simply id is used when A is clear from the context.

   The morphisms must satisfy the following axioms:

   (i) $f : A \rightarrow B \wedge f : A_0 \rightarrow B_0 \Rightarrow A = A_0 \wedge B = B_0$ (unique type),
   (ii) $f : A \rightarrow B \wedge g : B \rightarrow C \Rightarrow f; g : A \rightarrow C$ (composition type),
   (iii) $idA : A \rightarrow A$ (identity type),
   (iv) $(f; g); h = f; (g; h)$ (composition),
   (v) $id; f = f = f; id$ (identity).

   Note that $f : A \rightarrow A$ does not mean that f is an identity (it is an *endomorphism*). For example, if X is the space of reals, the function $y(x) = x^3$ maps X into itself, but it is not an identity. Also $f : A \rightarrow B$ and $g : A \rightarrow B$ does not mean that $f = g$. Sometimes there are several different categories under discussion. Then the name of the category must be added to the above notations as a subscript in order to avoid ambiguity.

   Let $\mathcal{A}$ and $\mathcal{B}$ be categories. Then a *functor* from $\mathcal{A}$ to $\mathcal{B}$ is a mapping F that sends objects of $\mathcal{A}$ to objects of $\mathcal{B}$ and morphisms of $\mathcal{A}$ to morphisms of $\mathcal{B}$ in such a way that:

   (vi) $Ff : FA \rightarrow_{\mathcal{B}} FB$ whenever $f : A \rightarrow_{\mathcal{A}} B$,
   (vii) $Fid_A = id_{FA}$ for each object A in $\mathcal{A}$,
   (viii) $F(f; g) = Ff; Fg$.

   The *operation of modeling* (creating a model) can be represented by a functor in the categorical language. This consists in mapping the objects and the interaction rules from the real world to our minds or to the mathematical description or computer software.

   A useful concept in the category theory is dualism. The dual of a term in the categorical language is defined as follows:

dualA = *A* for object *A*,
dual(*f* : *A* → *B*) = *dualf* : *B* → *A* (note the swap of *A* and *B*)
dual(*f*; *g*) = *dualg*; *dualf* (note the swap of *f* and *g*),
dual(*idA*) = *idA*.

Clearly, *dualizing* is its own inverse, that is, *dual*(*dual y*) = *y* for each term *y*. For each definition of a concept or construction xxx expressed in the categorical language, we obtain another concept, often called co-xxx (if no better name suggests itself), by dualizing each term in the definition (Fokkinga [13]). Also, for each equation *f* = *g* provable from the axioms of category theory (and hence valid for all categories), the equation *dualf* = *dualg* is provable too. Thus, dualization cuts work in half, and gives each time two concepts or theorems for the price of one.

Here, we use the category language in a "non-mathematical" mode, i.e., to manage models and not the mathematical theories. Consider the following ODE model:

$$\frac{dx}{dt} = f(x, t), \ x(0) = x_0. \tag{1.27}$$

To use the categorical language, let us define the category object as a pair $(x(t), t)$, where $x(t)$ is the system state. Suppose that a morphism exists between two objects $(x(t_a), t_a)$ and $(x(t_b), t_b)$ if and only if $t_a \leq t_b$. A morphism g between these two objects can be defined as "solve Eq. (1.27) over $[t_a, t_b]$ with the initial condition $x(t_a)$." Note that this means that in the model (1.27), the time cannot go backwards. It is rather a philosophical question if the real world is reversible in this sense. The common believe is that it is not. Formally, using the model (1.27), we could integrate the equations backwards to restore the previous states (with certain accuracy). However, this is not so simple in the case of the following model (a DI with uncertainty represented by the right-hand side set-valued function):

$$\frac{dx}{dt} \in F(x, t), \ x(0) \in X_0. \tag{1.28}$$

Observe also that the morphism g may be defined as any other well-defined state transition algorithm, not necessarily a differential equation or inclusion. This may be some logical procedure with discrete state and discrete time. In other words, this categorical specification can also be applied to discrete event models because it does not use the internal structure of objects.

1.7. The inverse problem

The problem for an ODE model is: Given the final state of a dynamic system, what was the initial state that originated it?

For the ODE model (1.27) we can simply go backwards in time, from t_1 to t_0. This means that we resolve the equation

$$\frac{dx}{ds} = -f(x, t), \tag{1.29}$$

with initial condition $x(t_1)$, over an interval $I = [t_1, t_0]$ (inverse time axis). In this case, the inverse problem is trivial.

Now, consider a similar inverse problem for the model (1.28) (a dynamic system with uncertainty): Given the final reachable set, what was the initial set that originated it? In this case we cannot just go backwards in time. Indeed, consider, for example, the inclusion

$$\frac{dx}{ds} \in [-1, 1], \; x \in R. \tag{1.30}$$

Starting from the initial point $x = 0$, $t = 0$ and calculating the reachable set at $t = 1$, we get $RS(0, 1) = [-1, 1]$. Now, if we start from $t = 1$ with the final set $[-1, 1]$ and go backwards (inverting the sign of the derivative), we get the set (interval) $[-2, 2]$ instead of the original set (point) $(0, 0)$. It is clear that the reachable set must grow because the uncertainty remains the same for the inverse problem. For the reader convenience, we repeat here Fig. 1.7, see Fig. 1.9. The figure illustrates this in the 2D case. The reachable set obtained by inverting the sign of the derivatives (going backwards in time) is denoted as BRS. In other words, note that while the solutions to ODEs are reversible (supposing an analytic solution, or ideal, error-free numeric method), the solutions to DIs are not.

In the general case of the DI (1.28) over a fixed time interval, the solution to the inverse problem should result in the initial set A, (see Fig. 1.9). Denote this solution by IRS. Thus, we should have $IRS(RS(A, t1), t0) = A$. In other words, we must be able to express the set A in terms of the reachable set at $t = t_1$. The second parameter of IRS (as well as of BRS) is the target time instant.

In [32] it has been pointed out that such defined IRS is the intersection of all BRSs (or their closures) going out of the points of B as initial sets (at t_1, backwards in time). We will not repeat here the proof given in that article because the present book is focused on practical applications and numerical examples.

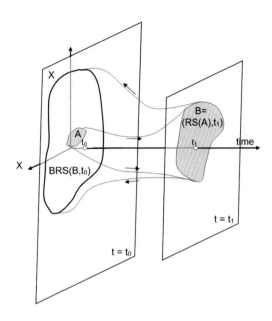

Figure 1.9 Initial set *A*, the reachable set (RS), and the backward reachable set (BRS).

One may conclude that the above considerations mean that dynamic systems with uncertainty are reversible. However, from the computational point of view this is hardly true. The reversing procedure includes calculation of the intersection of all reachable sets of points of the final reachable set, which means infinite reachable set calculations (infinite computational complexity). In other words, finding the inverse reachable set is a task of higher computational complexity and may result to be intractable.

1.8. The discrete version

Roughly speaking, computer simulation, as well as model building, has two main branches: continuous simulation and discrete event simulation. Some models may contain both discrete and continuous parts, known as combined models. In this book, we deal with continuous modeling simply because the DI is a generalization of the differential equation and implies continuity of the involved functions and other regularity assumptions. However, observe that the general concept of transforming the model state in time can be considered in the continuous, as well as the discrete case. By the *discrete case* we understand both time and state space discretization.

In such case the name *DI* can no longer be used because of its relation to differential equations. We will discuss the term *discrete decision inclusion* (DDI) rather than the (discrete) DI.

This approach can always be applied when we consider a system where the participating model components advance in time, changing from one discrete state to another. This change is instantaneous, and time advances by a given interval. To select the next state, the component must make a decision "where to go" and then jump to the selected state. This decision may be unique and well defined, random, or uncertain, belonging to a set of permissible and limited decisions. The latter case is the discrete version of the DI.

In the discrete case, we make an important assumption: The steps (increments) of time as well as of the model state are *always finite*. *No operations* or considerations that involve *passing with these increments to zero* and looking for some limit values or properties are considered.

The present section does not contain any new theoretical considerations, and should only be treated as a suggestion for further research based on some analogies between DIs and DDIs.

Perhaps, DDIs can be treated as some kind of cellular automata. Recall that a cellular automaton consists of a collection of cells on a grid in space. Each cell can evolve through consecutive time steps, due to given rules of change. These rules may be constant or variable in time, and normally depend on the state of the cell and its neighborhood. Cellular automata evolve, creating interesting patterns and sometimes quite unexpected images. This pattern creation and cluster moving resembles biological systems and is sometimes called "artificial life." Cellular automata are not subject of this book, so we will not discuss them here. Consult, for example, Von Neumann [50,49] or Wolfram [47].

However, the DDI approach is not exactly a cellular automaton. In cellular automata, each cell has its state. While advancing in time, the cell may change its state, for example from "0" to "1." In a DDI, the cells represent possible states of the model, and the model trajectory jumps from one to another, changing the state of the model and not of the cells.

In a DDI, we have the following elements: the space of states, 1D or multidimensional grids of states, the initial set the model movement starts from, and the rules of change and additional state restrictions. Compared with continuous DIs, the problem statement is similar. The rules of change of a DDI correspond to the right-hand side of the DI (in the continuous case it is the set where the derivative of motion must belong). The DDIs

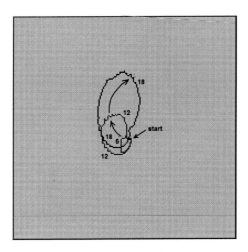

Figure 1.10 Evolution of the boundary of the reachable set. Model 1.

may be a useful tool in system simulation and decision making. Any trajectory of a DDI is a sequence of states defined by the decisions taken about the system movement. If we associate each state with an object function (e.g., a cost function), then we may consider an optimization problem for a decision-making system. Here, we will not deal with any detailed and theoretical issues. Let us only show some images of the simulated reachable sets that could give a useful view on the DDI dynamics through a graphical representation of results.

Consider a DDI whose states are represented by cells in 2D grids of states. The initial condition is given as one cell where the model state belongs. Model states are defined on a square area with dimensions 200×200 cells. The initial cell is placed at point $(5, 7)$. The vertical axis is directed downwards.

1.8.1 Model 1: a linear rule

The rule of state change is as follows. The horizontal (x) increment is equal to $(j - m)/6 + m/4$, the vertical increment is equal to $(k - n - j + m)/4$, and k and j scan the values in the ellipse of vertical radius 6 and horizontal radius 3 around the cell position (n, m). Thus, the parameters k and j define the model uncertainty.

Fig. 1.10 shows the evolution of the boundary of the reachable set after 6, 12, and 18 time steps. In each time step, the new position is the actual position plus the corresponding increments. The 3D image of the reachable

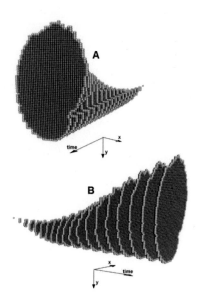

Figure 1.11 3D images of the reachable set of model 1.

set can be seen in Fig. 1.11. Each sphere is an attainable state. The white spheres mark the boundary of the reachable set. Panels A and B depict the reachable set viewed from two different angles.

1.8.2 Model 2: non-linear model

This is a non-linear model. The rule of state change is as follows. The horizontal (x) increment is equal to $(k - n + m)^2/80$ and the vertical increment is equal to $(n + m - j)/3$. The point (k, j) scans a circle of radius 3 around the current point (n, m). Fig. 1.12 shows the 3D image of the reachable set.

1.8.3 Model 3: state variable restrictions

This model is presented to show the effects of state variable restrictions. The change of state rule is as follows. The horizontal increment is equal to $(j.m)/6 + m/5$ and the vertical increment is equal to $(k - n - j + m)/4$. The point (k, j) scans the circle of radius 6 around the current position (n, m). The prohibited state variables area is a circle of radius 10 around the point $(20, 0)$ (see Fig. 1.13).

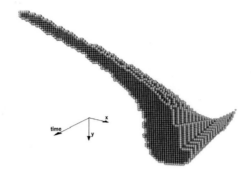

Figure 1.12 The reachable set for model 2.

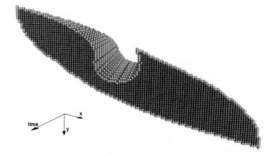

Figure 1.13 The reachable set of a DDI with state restrictions. Model 3.

1.9. Conclusion

Some basic definitions related to DIs have been recalled in this chapter. The notion of quasitrajectory has been explained. The relation to control systems is an important characteristic of DIs. This relation and the properties of reachable sets play an important role in optimal control theory. The application of DIs in uncertainty treatment, as well as the term *solution to the DI* have been discussed. We highlight that the trajectories of a DI should not be treated as solutions. It is more reasonable to treat the reachable set as the solution. However, this is not a common definition of the solution. Thus, we avoid the term solution, using the notion of reachable set instead.

The concept of functional sensitivity has been discussed. It was pointed out that this kind of sensitivity is closely related to DIs and the reachable set can be treated as the solution to the problem.

While ODE models are reversible (they can be integrated backwards in time), DI models are not. We gave a possible generalization of the model in categorical language, which is much more abstract and can work for continuous, as well as discrete event models. The categorical specification of both kinds of models is the same. Using the concept of dualism, we can construct the dual category to solve the inverse problem. In the case of models with uncertainty (which always exists in the real world), the inverse problem cannot be solved by backward solution of the corresponding DI. The proposed procedure for the inverse problem (dualizing the original morphisms) is not very complicated, but may be difficult when looking for numerical implementations.

The main topic of this book is the use of DIs to deal with the uncertainty in dynamical systems. It is difficult to compare this approach to commonly used stochastic methods. The reachable set of a DI is a deterministic object, while models with stochastic variables provide information about the probabilistic properties of trajectories with uncertainty. We cannot say which method is better in uncertainty treatment. This depends on each particular case, on the input data (parameter limits or probability distributions), and on the purpose of modeling and simulation defined by the user.

Further research should also be done on the discrete version, called here DDI. Possible analogies with continuous DIs should be investigated.

References

[1] A. Asperti, G. Longo, Categories, Types and Structures, MIT Press, ISBN 0262011255, 1991, ftp://ftp.di.ens.fr/pub/users/longo/CategTypesStructures/book.pdf.

[2] J.P. Aubin, L. Chen, O. Dordan, Tychastic Measure of Viability Risk, Springer International Publishing, ISBN 978-3-319-08128-1, 2014.

[3] J.P. Aubin, P. Saint-Pierre, A tychastic approach to guaranteed pricing and management of portfolios under transaction constraints, in: Seminar on Stochastic Analysis, Random Fields and Applications V, Centro Stefano Franscini, Ascona, 2005.

[4] J.P. Aubin, A. Cellina, Differential Inclusions, Springer Verlag, Berlin, ISBN 978-3-642-69514-8, 1984.

[5] L.D. Berkovitz, Variational Approach to Differential Games, Princeton Univ. Press, 1964.

[6] D.G. Cacuchi, Sensitivity and Uncertainty Analysis: Theory. I, Chapman & Hall, ISBN 1584881151, 2003.

[7] F. Campolongo, A. Saltielli, Sensitivity analysis of an environmental model: an application of different analysis methods, Reliability Engineering & Systems Safety 57 (1) (1997) 49–69, https://doi.org/10.1016/S0951-8320(97)00021-5.

[8] R.D. Cook, S. Weisberg, Criticism and influence analysis in regression, Sociological Methodology 13 (1982) 313–461.

[9] K.E. Dutton, Optimal Control theory Determination of Feasible Return-to-Launch-Site Aborts for the HL-20 Personnel Launch System Vehicle, Report NASA Technical Paper NAS 1.60:3449, 1994.

[10] S. Eilenberg, Automata, Languages, and Machines. 1, 2, Academic Press, New York, 1974.

[11] L.D. Elsgolc, Calculus of Variations, ISBN 978-0486457994, 2007.

[12] A.F. Filippov, Classical solutions of differential equations with multivalued right hand, SIAM Journal on Control 5 (1967) 609–621.

[13] M.M. Fokkinga, A gentle introduction to category theory, Internet communication, 1994, http://wwwhome.cs.utwente.nl/~fokkinga/mmf92b.pdf.

[14] M.M. Fokkinga, Law and Order in Algorithmics, Thesis/dissertation, University of Twente, Dept. Comp. Sc., 1992.

[15] D.A. Freedman, Statistical Models: Theory and Practice, Cambridge University Press, 2005.

[16] M. Friendly, D. Dennis, The early origins and development of the scatterplot, Journal of the History of the Behavioral Sciences 41 (2) (2005) 103–130, https://doi.org/10.1002/jhbs.20078.

[17] D. Fudenberg, J. Tirole, Game Theory, MIT Press, Cambridge MA, 1991.

[18] J. Gomez, R. Sanz, Modeling cognitive systems with Category Theory, Report, Universidad Tecnica de Madrid, 2009, http://tierra.aslab.upm.es/documents/controlled/ASLAB-A-2009-023.pdf.

[19] S.V. Grigorieva, V.N. Ushakov, Use finite family of multivalued maps for constructing stable absorption operator, Topological Methods in Nonlinear Analysis 50 (1) (2000).

[20] P.R. Halmos, Naive Set Theory, van Nostrand Company, Princeton, NJ, 1960.

[21] C.A.R. Hoare, Notes on an approach to category theory for computer scientists, in: M. Broy (Ed.), Constructive Methods in Computing Science, in: NATO Advanced Science Institute Series. Series F: Computer and System Sciences, vol. 55, Springer-Verlag, 1989.

[22] R. Isaacs, Differential Games, Dover Publications, Inc., New York, ISBN 9780486406824, 1999.

[23] P. Krbec, On nonparasite solutions, Internet communication, in: Equadiff 6, Brno J.E. Purkyne University, Department of Mathematics, 1986, pp. 133–139, http://eudml.org/doc/220003.

[24] E.B. Lee, L. Markus, Foundations of Optimal Control Theory, Wiley, New York, ISBN 978-0898748079, 1967.

[25] A. Marchaud, Sur les champs de demi-cones et les équations differielles du premier ordre, Bulletin de la Société Mathématique de France 62 (1934).

[26] B.S. Mordukhovich, Optimal Control of Nonconvex Differential Inclusions, Report, Institute for Applied Systems Analysis, 1997, http://pure.iiasa.ac.at/5261.

[27] M.D. Morris, Factorial sampling plans for preliminary computational experiments, Technometrics 33 (1991) 161–174, https://doi.org/10.2307/1269043.

[28] J. Nearing, Mathematical Tools for Physics, Dover Publications, 2010.

[29] L. Petrosjan, N.A. Zenkiewicz, Game Theory, World Scientific Publishing Co., Inc., 1996.

[30] A. Plis, Remark on measurable set-valued functions, Bulletin de L'Académie Polonaise Des Sciences. Série Des Sciences Mathématiques, Astronomiques Et Physiques 9 (12) (1961) 857–859.

[31] L.S. Pontryagin, The Mathematical Theory of Optimal Processes, Wiley Interscience, New York, 1962.

[32] S. Raczynski, Uncertainty, dualism and inverse reachable sets, International Journal of Simulation Modelling (ISSN 1726-4529) 10 (1) (2011) 38–45.

[33] S. Raczynski, Some remarks on nonconvex optimal control, Journal of Mathematical Analysis and Applications 118 (1) (1986) 24–37, https://doi.org/10.1016/0022-247X(86)90287-8.

[34] S. Raczynski, On some generalization of "bang-bang" control, Journal of Mathematical Analysis and Applications 98 (1) (1984) 282–295, https://doi.org/10.1016/0022-247X(84)90295-6.

[35] Saunders Mac Lane, Categories for the Working Mathematician, Graduate Texts in Mathematics, Springer, ISBN 0-387-98403-8, 1998.

[36] R. Sentis, Equations diferentielles a second membre mesurable, Bollettino dell'Unione Matematica Italiana (ISSN 1972-6724) 15 (B) (1978) 724–742.

[37] I. Sobol, Sensitivity analysis for non-linear mathematical models, Mathematical Modeling and Computational Experiment 1 (1993) 407–414.

[38] E. Solan, N. Vielle, Quitting games, Mathematics of Operations Research 26 (2) (2001) 265–285.

[39] K. Sriyudthsak, H. Uno, R. Gunawan, F. Shiraishi, Using dynamic sensitivities to characterize metabolic reaction systems, Mathematical Biosciences 269 (2015) 153–163.

[40] S. Tsunumi, K. Mino, Nonlinear strategies in dynamic duopolistic competition with sticky prices, Journal of Economic Theory 52 (1990) 136–161.

[41] A. Turowicz, Sur les zones d'emision des trajectoires et des quasitrajectoires des systemes de commande nonlineaires, Bulletin de L'Académie Polonaise Des Sciences. Série Des Sciences Mathématiques, Astronomiques Et Physiques 11 (2) (1963).

[42] A. Turowicz, Sur les trajectoires et quasitrajectoires des systemes de commande non-lineaires, Bulletin de L'Académie Polonaise Des Sciences. Série Des Sciences Mathématiques, Astronomiques Et Physiques 10 (10) (1962).

[43] T. Wazewski, On an optimal control problem differential equations and their applications, in: Conference Paper: Proceedings of the Conference Held in Prague, Publishing House of the Czechoslovak Academy of Sciences, Prague, 1963.

[44] T. Wazewski, Sur une genralisation de la notion des solutions d'une equation au contingent, Bulletin de L'Académie Polonaise Des Sciences. Série Des Sciences Mathématiques, Astronomiques Et Physiques 10 (1) (1962).

[45] T. Wazewski, Sur les systemes de commande non lineaires dont le contredomaine de commande n'est pas forcement convexe, Bulletin de L'Académie Polonaise Des Sciences. Série Des Sciences Mathématiques, Astronomiques Et Physiques 10 (1) (1962).

[46] T. Wazewski, Sur une condition equivalente a l'equation au contingent, Bulletin de L'Académie Polonaise Des Sciences. Série Des Sciences Mathématiques, Astronomiques Et Physiques 9 (12) (1961).

[47] S. Wolfram, Universality and complexity in cellular automata, Physica D. Nonlinear Phenomena 10 (1–2) (1984) 1–35, https://doi.org/10.1016/0167-2789(84)90245-8.

[48] S.K. Zaremba, Sur les équations au paratingent, Bulletin Des Sciences Mathématiques 60 (1936).

[49] J. von Neumann, A.W. Burks, Theory of Self-Reproducing Automata, University of Illinois Press, 1966.

[50] J. von Neumann, The general and logical theory of automata, in: Cerebral Mechanisms in Behavior - the Hixon Symposium, John Wiley & Sons, New York, 1951.

CHAPTER TWO

Differential inclusion solver

Abstract

Solving a differential inclusion (DI) is more difficult that solving an ordinary differential equation. First, we should define the term "solution to a DI." As there is some confusion about this, we will avoid the word "solution" (commonly used for a single DI trajectory) and rather discuss *reachable sets*. The aim of the DI solver described below is to obtain and to show graphically the reachable sets of DIs. The algorithm is based on some concepts from optimal control theory. As mentioned before, information about the reachable set may be useful when dealing with uncertainty in model parameters and when analyzing functional sensitivity. A common error encountered in many algorithms to compute reachable sets is that the interior of the set is explored. The DI solver described here scans the *boundary* and not the *interior* of the set. The solver may also be used in the analysis of control systems.

Keywords

Differential inclusion, DI solver, Reachable set

2.1. Main concepts

The definitions and properties of differential inclusions (DIs) have been given in Chapter 1 from a more mathematical point of view. Here, we will discuss some more practical issues that result in a computational procedure named *DI solver*. This procedure calculates the shape of reachable sets of DIs, as well as of control systems. The original version of the solver was published in Raczynski [11].

Consider a simple example of a second-order system,

$$a\frac{d^2y}{dt^2} + b\frac{dy}{dt} + y(t) = 1. \tag{2.1}$$

This is an ordinary differential equation (ODE) model. Introducing the notation $x_1 = y$, $x_2 = dy/dt$, we obtain

$$dx_1/dt = x_2, \quad dx_2/dt = (1 - bx_2 - x_1)/a. \tag{2.2}$$

In more general notation, the state equation is

$$x'(t) = f(a, b, x), \quad f_1 = x_2, \quad f_2 = (1 - bx_2 - x_1)/a, \tag{2.3}$$

Reachable Sets of Dynamic Systems
https://doi.org/10.1016/B978-0-44-313384-8.00019-1

where x is a 2D vector, $x = (x_1, x_2)$, t is the time, and f is a vector-valued function. The prime sign means time differentiation.

Now, suppose that the parameters a and b are uncertain and that the only information we have are the corresponding intervals where their values can belong or a permissible set (which may be quite irregular and variable) on the plane where the point (a, b) must be located. Moreover, we treat a and b as variables, so they are not fixed along the system trajectory and may change in time. Note that we know nothing about a possible probability distribution of these parameters and we do not treat them as random variables. Thus, Eq. (2.3) takes the following form:

$$x' \in F(x, t), \quad x(0) \in X_0, \tag{2.4}$$

where F is a set (all possible values of f, while a and b scan their permissible set). X_0 is the initial set that may reduce to one point. What we obtained is a DI. The right-hand side of Eq. (2.3) determines the set F. In this case, it is parametrized by a and b. The state vector $x \in R^n$, and F is a mapping from $R^n \times R$ to subsets of R^n (the real n-dimensional space). More detailed definitions and assumptions have been given in Chapter 1.

DIs are closely related to control systems (Wazewski [16]). To see this, consider a dynamic control system

$$x' = f(x(t), u(t), t), \quad x(0) = x_0, \quad u(t) \in C(x(t), t) \subset R^m, \tag{2.5}$$

where $u(t)$ is the control variable. We can define the corresponding mapping F as follows:

$$F(x, t) = \{z : z = f(x, u, t) \mid u \in C(x, t)\}, \tag{2.6}$$

where the set C represents the restrictions for the control variable. Using this mapping in (2.4), we get a DI model of the control system (2.5), where the controls do not appear explicitly. This relation to control systems makes DIs useful to those who work in the field of control theory, in particular in optimal control (Wazewski [16,17,15]). When dealing with DIs, the main topics considered are the topological properties of the set $Z(x(0))$, which is the union of the graphs of all trajectories of the DI with initial condition $x(0)$. This set is called *emission zone* or *reachable* or *attainable set* of the DI. Consult Chapter 1 for more details on DIs.

The reachable set for the possible system trajectories may be used when dealing with the uncertainty problem. In this very natural way, we see that the uncertainty in dynamic system models leads to DIs as the corresponding mathematical tool (see Section 1.4 for more comments). This tool has been known for about 70 years, and there is a large amount of literature available on DI theory and applications. The first works were published in 1934–1936 by Marchaud [7] and Zaremba [18].

Most of the publications on solutions of DIs deal with problems of existence. Note that in all those publications, by "solution" the authors mean a DI trajectory. Recall that in this book we do not treat a trajectory as a solution. Here, the (possibly unique) solution to a DI is the corresponding reachable set. Thus, we do not use the term "solution" to avoid misinterpretations. Bressman [2,1] considers the existence of trajectories and their properties.

Carja [3] discusses problems of flow invariance, stability, and control theory of DIs and regularity properties of the solutions. Filippov and Plis provide more properties of solutions to DIs [5,8]. The connection to control systems is also mentioned in that paper, as well as issues of DIs in Banach spaces. Collins and Graca [4] deal with problems of computability.

2.2. Solver algorithm

Someone working with differential equations can find a huge number of numerical methods and software programs. However, for DIs there are few tools that could help a simulationist. The *DI solver* described here is the result of an attempt to fill this gap. Note that the original algorithm of the solver was already published in 2002 by Raczynski [11].

A common error in reachable set determination algorithms is that the inside of the set is explored, looking for a uniform distribution of interior points. The algorithm proposed here explores the boundary and not the interior of the set.

The software described below is not a commercial package. The source code is available from the author on request. However, note that the code is hardly portable and may be difficult to run in software environments other than the one used here. The solver algorithm has been coded in Embarcadero Delphi and requires that package to be installed on the user's machine. Another way to implement the solver may be to code it from the very beginning, using the algorithm described below. A limited, standalone ".exe" version of the solver is also available, but it is considerably slower and limited. It should be emphasized that our main goal is the determination of the reachable set, and not optimization. The DI solver and the present problem statement should not be confused with the DI method used in optimal control problems.

As stated before, we are looking for the reachable set with a given initial condition or initial set. It is known that with sufficient regularity assumptions (mainly the continuity of the convex field E and the generalized Lipschitz condition, see Section 1.2.2), the reachable set is continuous with respect to the initial condition (Turowicz [13,14]) and it is a connected set for any fixed time instant. However, even if $F(x, t)$ is always convex, the reachable set need not be convex and may have a complicated shape. It might appear that a simple way to obtain the shape of the reachable set is to calculate a number of different trajectories integrating the equation

$$\frac{dx}{dt} = z(t),$$

where $z(t)$ is a selector (not a trajectory) of the DI, that is, $z(t) \in F(x(t), t)$ a.e. over a given time interval I. It might be expected that choosing $z(t)$ randomly from inside of F, we can cover the inside of the reachable set with a density that is sufficient to estimate its shape. One could expect that if the selections are made according to a uniform probability density inside F, then the shape of the reachable set can be estimated. Unfortunately, this is not the case, even if we select only points from the boundary of the set F. We will call this method *primitive shooting* and compare it with the DI solver algorithm. Simulation experiments show that even in very simple cases, the set of trajectories provided by primitive shooting (using any density function) is concentrated in some small region inside the reachable set and does not approach its boundary. The problem is that the resulting set of trajectories has a very narrow distribution inside the reachable set, even if the distribution used to generate selectors is not narrow, for example uniform inside F.

Our method consists in random search of the boundary of the reachable set, and not of its interior. As the result, we obtain a set of points that belong to the boundary. With enough points, the shape of the reachable set can be estimated with reasonable accuracy. The problem is to generate these trajectories in such a way that the density of points is nearly uniform on the boundary. This permits to avoid "holes" in the resulting graphical image.

The mechanism of generating "boundary trajectories" is well known in control theory. If the set F is convex, then the graph of any trajectory that passes through a boundary point of the reachable set at $t = t_1$ must entirely belong to the boundary for all $t < t_1$ (Zaremba [18] and Lee and Markus [6]). On the other hand, each boundary trajectory is the optimal solution to a certain dynamic optimization problem, and can be calculated using, for example, the maximum principle of Pontryagin [10]. If the field is not convex, but Lipschitzian, it is sufficient to estimate the reachable set for trajectories of the corresponding tendor field (see Chapter 1) instead of the original field F. This can be easier because the tendor field contains few points, in many cases only isolated extremal points of the given set F. Recall that the fields F, E, and Q (Fig. 1.3 in Chapter 1) have the same quasitrajectories and that for each quasitrajectory a regular trajectory exists nearby.

The maximum principle of Pontryagin [10] states that a necessary condition for a trajectory to be optimal is that the *Hamiltonian* for each time instant over the time interval under consideration is being maximized (see Chapter 3). In other words, the maximum principle permits us to decompose the original optimization problem of maximization of a functional into a set of problems of function maximization. The original optimization problem is as follows. Given a control system described by Eq. (2.5), we look for an optimal control and the corresponding optimal trajectory that minimizes a given criterion J over the interval $I = [0, T]$, with J defined as

$$J = \int_0^T f_0(t)\,dt, \qquad (2.7)$$

where f_0 is the function that defines the optimization criterion. If $f_0 \equiv 1$, then the trajectory is time-optimal, i.e., it reaches the final point in optimal time. To define the Hamiltonian, we must define the *conjugated vector* $p \in R^n$ that satisfies (by definition) the following equations:

$$p_i' = -\sum_{j=1}^{n} \frac{\partial f_j}{\partial x_i} p_j - \frac{\partial f_0}{\partial x_i}, \qquad (2.8)$$

where $i = 1, .., n$ and f is the vector of the right-hand sides of (2.5). Observe that a necessary condition for a trajectory to be time-optimal is to entirely lay on the boundary of the reachable set. This means that we can suppose $f_0 \equiv 1$ and eliminate f_0 from the above equations.

The Hamiltonian is defined as follows:

$$H = \sum_{j=1}^{n} p_i f_j. \qquad (2.9)$$

According to the maximum principle, the necessary condition for the trajectory to terminate at a boundary point of the reachable set is that the control $u(t)$ maximizes the Hamiltonian over the interval I. This can be used to generate boundary trajectories of the DI. If the inclusion is given in the form of a control system (2.5), then we apply the principle directly. If it is given in the general form (2.4), we must parametrize the set F and treat the parameter as the control.

It is known from optimal control theory that at a given final point of a boundary trajectory, the final value of the vector p must satisfy the *transversality conditions* (Lee and Markus [6]). This provides the final condition for the conjugated vector. Consequently, to calculate the optimal trajectory with a given optimality criterion, we must solve the *two-point boundary value problem*. The boundary conditions for the state vector are given at the initial time, and the boundary conditions for the conjugated vector are known at the end of the trajectory. A common way to solve this problem is to apply a certain iterative algorithm that approaches the desired final state. Such algorithms need multiple forward and backward integrations of the set of Eqs. (2.5) and (2.8).

In our case, no backward iterative integrations are needed. Observe that starting with *any* initial condition for the conjugated vector and maximizing the Hamiltonian in each time step, a single forward integration of Eqs. (2.5) and (2.8) provides a trajectory that scans the boundary of the reachable set. Thus, we can choose the initial conjugate vector randomly, obtaining random final boundary points (and not the points inside the reachable set). We can start with any (e.g., random) initial condition because we do not solve any particular optimization problem, but we are just looking for the trajectories that scan the boundary of the reachable set. The problem is how to generate the initial vector p to cover the resulting final boundary set with uniform density of points and to avoid "holes" in it.

The distribution D used below is the probability distribution function defined inside the n-dimensional unit cube with center at the origin of the coordinate system. The algorithm is as follows (the discrete-time version of the maximum principle is used).

DI solver algorithm

0. Set D equal to the uniform density function, set $x = x_0$.
1. Generate initial vector p according to the density D.
3. Integrate Eqs. (2.5) and (2.8) over the interval $I = [0, T]$ using the control that maximizes the Hamiltonian at each integration step.

4. Store the initial p and the whole trajectory in a consecutive record of a file.

5. Select the final point x_k that lies in a region of minimal density of points $x_i(T)$, searching in the file where trajectories have been stored.

6. Modify the distribution D, increasing the probability density in a neighborhood of the point p_k that corresponds to the point x_k.

7. If there are enough trajectories stored, then stop; otherwise, go to step 1.

By the term "region of minimal density of points" (point 5) we understand the spot of low density of points on the image of the final reachable set or on its 2D projection. The simplest way to find it is to scan all the points and to select a point that has the maximal distance from its nearest neighbors. This part of the algorithm is rather heuristic. In fact, starting the algorithm, we do not know the real shape of the reachable set (this is what we are looking for).

Step 7 of the algorithm contains a stop condition, related to the number of trajectories. This number cannot be exactly defined, so the stop condition may rather be based on the user's decision. As the result, we obtain a 2D image that is a projection of the boundary of the reachable set into a given plane. In the case of a second-order system this image should be a closed curve. For systems of higher order it will be a cloud of points in R^n. The user can stop the program if he/she can recognize the shape of the reachable set and identify its boundary. Practical experiments show that this can be reached after integrating 500–1000 system trajectories. Anyway, this procedure may be difficult to use for models of higher dimensionality (more than 10, perhaps).

The algorithm includes maximization of the Hamiltonian (point 3), which can be done using any maximization procedure. The program uses two user-defined procedures. One is the procedure that defines the right-hand sides of the system equations in (2.5) (the DI in parametrized form), and the other defines the algorithm for Hamiltonian maximization. These procedures are not predefined because they depend on each particular case (the model). Note that here we reduce the original problem of solving a DI to some sub-problems that may not be easy to solve, but have been tackled using well-known optimization techniques. If the original system is linear with respect to the control vector and the restriction set is a multidimensional cube, then the maximization can be reduced to a simple scan over a finite number of points. We will not discuss here the methods of maximization of the Hamiltonian (point 3). There exists a huge body of literature on it in the field of control theory (Lee and Markus [6] and Polak [9]).

To simplify the process of generating the initial conjugated vector, the density D is defined using marginal densities instead of a general n-dimensional density function. The program also contains another mechanism that helps us obtain a more uniform point distribution. This is a simple scanning that eliminates points which lie very close to each other or are identical. Such double points can appear because the system we analyze is a discrete-time approximation of the original continuous system. The effects of time discretization can be seen on the obtained images. There are some trajectories that are identical to each other, and certain points are not reachable due to the time discretization.

An important question is if and when you really need the reachable set calculated by the DI solver. Obviously, if our system is of the first order, the determination of the reachable set is trivial and can be done easily without involving DIs. In some cases, when the modeled system is of higher order, but strongly damped, the extreme points of the reachable set can be calculated simply by applying the extreme values of control variables. However, in a general multidimensional case (even if the system is of the second order and linear), the constant extreme controls (extreme points of the control restriction set) do not correspond to the extreme or boundary points of the system's reachable set. The reachable set assumes a complex multidimensional shape, not necessarily convex, with a boundary surface that may fold several times. Even in a simple 2D case and with an oscillatory model, the mapping from the permitted control set C to the reachable set may be highly irregular.

NOTE: The parametrization of the set F is an important and sometimes difficult task. Recall that while passing from the control system to the DI, we must map the tendor kernel of the set C into the tendor set Q of F (consult Sections 1.2.2 and 1.3). If the mappings are linear, we may suppose that the extremal points of C map into the extremal points of F. Unfortunately, in the general, non-linear case, this is not true. This means that the resulting reachable set may be incomplete. Note, however, that the algorithm cannot provide points located outside the reachable set because anyway, each integrated trajectory was calculated using the control that obeys restrictions C. Thus, any parametrization imperfection may lead to an estimate of the reachable set "from below."

2.2.1 Multiprocessing

The solver algorithm can be easily extended for machines with multiple processors. The only modification consists in calculating multiple trajectories concurrently. Steps 3 and 4 become as follows in the modified algorithm (NP is the number of processors).

Step 3M. Launch NP tasks of trajectory calculations according to (2.5) and (2.8) over the interval I using the control that maximizes the Hamiltonian at each integration step.

Step 4M. For each integrated trajectory, store the initial conditions for p, x and the whole trajectory in a consecutive record of a file.

This way, we accelerate the calculation approximately NP-fold.

2.2.2 Example 1: non-linear second-order model

Consider the following second-order non-linear model (the prime mark means time differentiation):

$$\begin{cases} x_1' = x_2, \\ x_2' = u - x_1 - 0.01x_2^3, \end{cases} \tag{2.10}$$

where the variable u fluctuates between -1 and 1. Fig. 2.1 shows a section of the reachable set of this system in the plane $t = 10$. In this figure we can compare the solver algorithm with primitive shooting. The points obtained by primitive shooting (10,000 trajectories) are concentrated in a small area and do not estimate the shape of the real reachable set. Fig. 2.2 shows a 3D image of the reachable set.

In this model, the damping term with x_2 is non-linear. Consequently, it can be observed that at the initial part the model oscillations grow, and then the damping maintains the model trajectories in a more stable region. The projections of the trajectories can be seen in Fig. 2.3. These are trajectories that scan the boundary of the reachable set. The oscillatory nature of those trajectories is clearly depicted. Not all trajectories can be seen in full. These calculated later obscure those integrated before.

2.2.3 Example 2: model with two uncertain parameters

Consider a second-order model with two uncertain parameters. We repeat here the equations mentioned earlier. The parameters are replaced by the components of control vector u, interpreted as uncertain variables:

$$dx_1/dt = x_2, \quad dx_2/dt = (1 - u_1x_2 - x_1)/u_2. \tag{2.11}$$

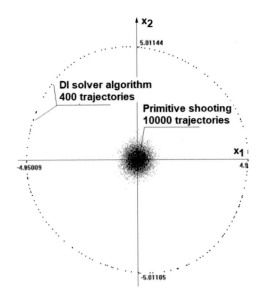

Figure 2.1 Time-section of the reachable set of Example 1.

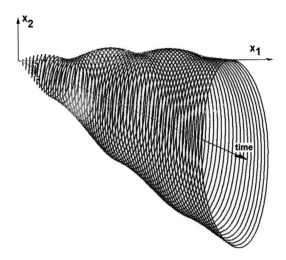

Figure 2.2 Example 1: 3D image of the reachable set.

Suppose that $u_1 \in [0, 1]$ and $u_2 \in [1, 2]$, changing their values along the model trajectory. The initial conditions are $x_1 = 0$ and $x_2 = 0$, and the final model time is equal to 13. Fig. 2.4 shows the intersection of the boundary of the reachable set with the plane $t = 12$, with about 400 trajectories

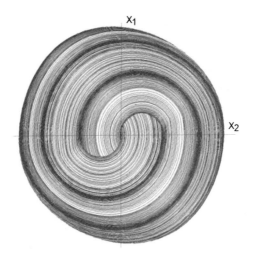

Figure 2.3 Example 1: Model trajectories projected on the $x_1 x_2$-plane.

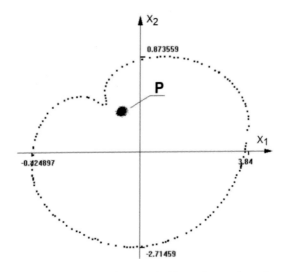

Figure 2.4 Example 2: Time section of the reachable set (500 trajectories stored). A cluster marked as P inside is the result of primitive random shooting (10,000 trajectories).

stored. Again, there is a small cluster of points (trajectory end points) obtained by primitive shooting with 10,000 trajectories. The cluster P of points in Fig. 2.4 shows how useless the simple primitive shooting is in the task of reachable set determination. Fig. 2.5 depicts the 3D image of the reachable set.

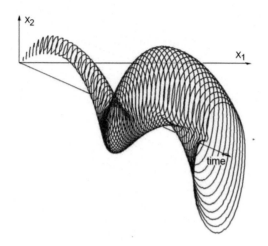

Figure 2.5 Example 3: 3D image of the reachable set.

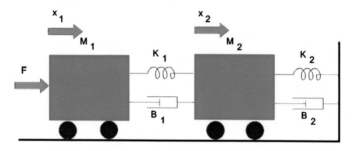

Figure 2.6 Example 3: A mechanical system.

2.2.4 Example 3: mechanical fourth-order systems

A simple mechanical system is shown in Fig. 2.6. The system equations are as follows:

$$\begin{cases} x_1' = x_3, \\ x_2' = x_4, \\ x_3' = [F - K_1(x_1 - x_2) - B_1 f(x_3 - x_4)]/M_1, \\ x_4' = [K_1(x_1 - x_2) + b_1 f(x_3 - x_4) - K_2 x_2 - B_2 f(x_4)]/M_2, \end{cases} \tag{2.12}$$

where $f(x) = x^2 sign\, x$. Note that the dampers are non-linear. F is an external force that belongs to $[-0.5, 0.5]$. The time section of the reachable set for this system is shown in Fig. 2.7. Here, time $= 4$, $m_1 = 1$, $m_2 = 2$, $k_1 = 0.3$, $k_2 = 0.1$, $b_1 = 1.5$, $b_2 = 3.0$. Note that some points appear to

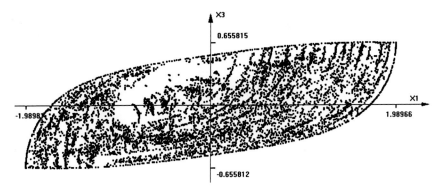

Figure 2.7 The shape of a time section of the reachable set of a fourth-order system (Example 3).

belong to the interior of the set. Those are not the points obtained by primitive shooting, as in the previous example. In fact, all these points belong to the boundary of the reachable set. What we see is a projection of a 4D figure (point cloud) onto a 2D plane x_1, x_2 for a given time instant. The graphical representation of the reachable sets for models of dimensionality greater than three is somewhat difficult to generate. The question is how to display a cloud of points of n-dimensional space in order to clearly show the shape. If the cloud is 3D, this can be done by rotating a 3D animated image to produce an illusion of 3D viewing. Other possible enhancements may be obtained using techniques known in fuzzy set theory. Fig. 2.8 shows the result of calculating the fuzzy variable representing the level of membership of the region. Points with membership values greater than 0.5 are shown as gray pixels. If there are not enough points to analyze, then the holes in the region disappear. Anyway, such images always depict approximate shapes.

2.2.5 Example 4: DI in general form

The models in the two previous examples were defined in the form of control systems. Now, let us consider a DI given in the general form of a 2D state vector.

The set F is shown in Fig. 2.9. It contains two disks of diameter 1 placed at $(-1, x_1)$ and $(1, -x_1)$, respectively. The two bold arcs indicate the corresponding tendor set.

Let us parametrize the set F. As stated in the previous section, we can use the tendor set instead of the original mapping F. This can be done as

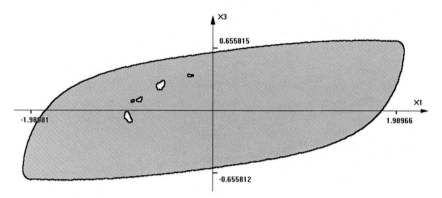

Figure 2.8 The shape of the reachable set of Fig. 2.7 enhanced by the fuzzy sets technique.

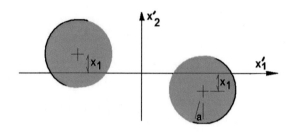

Figure 2.9 Example 4: The set F of the DI.

follows:

$$
\begin{cases}
x_1' = d_1(u), \\
x_2' = d_2(u, x), \text{ where } u \in [-1, 1], \\
d_1 = \begin{cases} -1 - 0.5cos(\pi(u + 1) - \pi/2 + a \text{ for } u < 0, \\ 1 - 0.5cos(\pi(u + 1 - \pi s/2 + a \text{ for } u \geq 0, \end{cases} \\
d_2 = \begin{cases} -x_1 - 0.5cos(\pi(u + 1) - \pi/2 + a \text{ for } u < 0, \\ x_1 - 0.5cos(\pi(u + 1 - \pi/2 + a \text{ for } u \geq 0, \end{cases} \quad \text{where } f(x) = x^2 sign\, x.
\end{cases}
$$

$$(2.13)$$

It can be verified that the point (d_1, d_2) scans the two parts of the tendor set when u changes from -1 to 1.

Fig. 2.10 shows the shape of the section of the reachable set of this DI with the plane $t = 3.5$. It is impossible to estimate this set by primitive shooting.

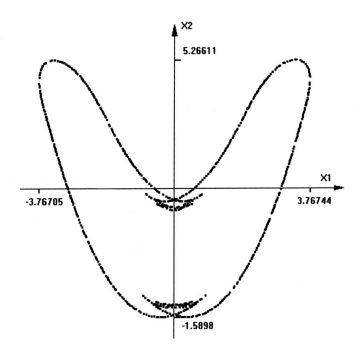

Figure 2.10 A time section for $t = 3.5$ (final time) of Example 4.

In some situations, when the boundary surface of the reachable set folds and intersects with itself, the algorithm may show some extra points inside the reachable set (Fig. 2.10). This may be due to the fact that we use a discrete-time model instead of the original continuous one. This can also be caused by imperfections in the Hamiltonian maximization procedure and by the fact that the maximum principle provides the necessary and not sufficient condition for optimality. Fig. 2.11 shows the 3D image of the reachable set of this example.

2.2.6 Example 5: Lotka–Volterra equations

Lotka–Volterra (L-V) equations describe the dynamics of ecological prey–predator systems (Takeuchi [12]).

In the simplest case of two species, the prey population (for example rabbits) grows due to the birth-and-death process. The growth would be exponential, but there is a limitation: There is a predator (e.g., wolves) who eat rabbits. The population of wolves grows when they have food, but if there are few rabbits available, the wolves die. We denote the rabbit

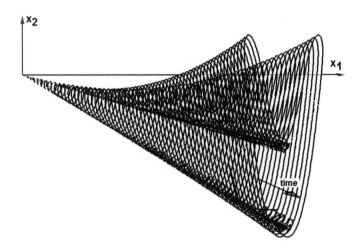

Figure 2.11 Example 4: 3D image of the reachable set.

population size as x_1 and the wolf population size as x_2. The classical form
of two-species L–V equations is as follows:

$$\begin{cases} x_1' = ax_1 - bx_1x_2, \\ x_2' = -cx_2 + dx_1x_2. \end{cases} \tag{2.14}$$

In the first equation, the term bx_1x_2 means that the rate of rabbits caught
by wolves is proportional to both the number of wolves and the number
of rabbits. A similar term appears in the second equation, telling that the
growth rate of wolves increases when they have more food. Coefficient a
defines the rabbits' natural birth rate, and c defines the wolves' natural death
rate. Many other versions of these equations are used in ecological models,
with two or more species in the n-species food chain.

Now, consider the two-species system with some uncertainty. Namely,
suppose that the birth rate of the rabbits is uncertain, subject to climate
changes and other external factors. For example, suppose that the parameter a may change in time, within the range of $\pm 25\%$. Thus, the system
equations can be written as follows:

$$\begin{cases} x_1' = a(1 + u)x_1 - bx_1x_2, \\ x_2' = -cx_2 + dx_1x_2, \end{cases} \tag{2.15}$$

where u changes between -0.25 and $+0.25$. This way we obtain a DI, with
the right-hand side parametrized by the variable u.

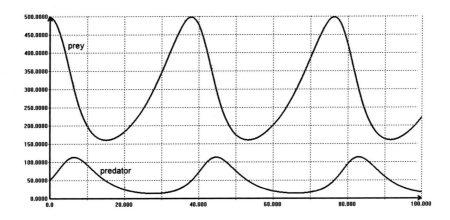

Figure 2.12 A simple simulation of the L-V model.

A simple simulation of the above equation with $u \equiv 0$ is shown in Fig. 2.12.

The L-V equations have strong non-linearities (the product of two state variables). Thus, the solution reveals a series of non-sinusoidal oscillations. The model parameters are $a = 0.1$, $b = 0.002$, $c = 0.3$, and $d = 0.001$, and the final simulation time is equal to 100.

Now, applying the DI solver we obtain the attainable set of the size of the two species in the presence of uncertainty. Fig. 2.13 shows the shape of the boundary of the reachable set for time $= 45$. In Fig. 2.14 we can see the 3D image of the set.

The projections of some boundary scanning trajectories are shown in Fig. 2.15. Fig. 2.16 shows similar trajectories with final time equal to 70.

In ecology and population growth models, we almost always have uncertain factors that have unknown probability distributions and stochastic parameters. In these cases, DIs may be a useful research tool. Observe that for the L-V model, even with small fluctuations of uncertain parameters, the size of the reachable set after time approximately equal to the model oscillation period is quite big. This means that this model is not very useful for predictions, even for small time intervals.

2.3. Conclusion

It is possible to create a versatile and flexible package for continuous- or discrete-time models described by DIs. On parallel computers, con-

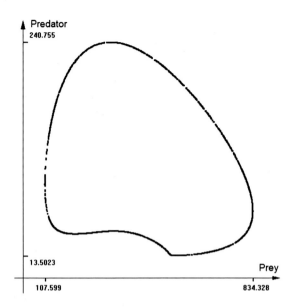

Figure 2.13 Time section of the reachable set for the L-V equations (time = 45).

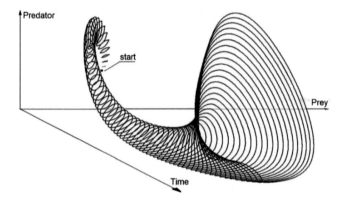

Figure 2.14 The shape of the reachable set of the L-V equation. 3D image.

current trajectory integration may be a natural and fast way to obtain the reachable sets. As in other methods and simulation packages, problems appear when dealing with stiff equations. For example, in models of order greater than three, the model may contain sub-systems that oscillate with very low and very high frequencies. In such cases the DI solver may fail. This issue needs further research.

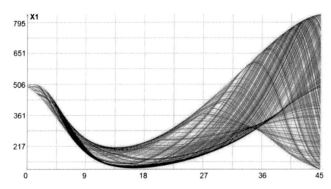

Figure 2.15 Some boundary scanning trajectories for Example 5. Projection on the x_1-time plane, with final time equal to 45.

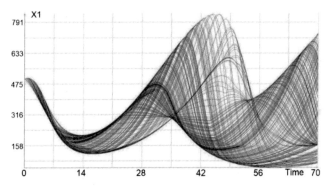

Figure 2.16 Some boundary scanning trajectories for Example 5. Projection on the x_1-time plane, with final time equal to 70.

Another important issue is the presentation of results. While solving differential equations, we can see the resulting trajectories as a set of plots, in function of time, or on the phase plane. The solution of a DI is a set rather than a curve. This implies a wide variety of possible graphical presentations of the solutions. The most general problem is how to show an n-dimensional non-convex set given by a cloud of points in the state space. Advanced graphical software and hardware already used in virtual reality programs can perhaps be adopted by more sophisticated DI solvers in the future.

DIs represent a powerful tool in uncertainty treatment. This may be uncertainty about the future, like the problem described in Chapter 11 "Uncertain future: a trip," as well as any other kind of uncertainty in dynamic systems. The main problems in developing DI solvers are the following.

1. Computational complexity. To advance one integration time step, the solver must perform a multidimensional optimization. Moreover, to estimate the shape of the DI solution (the reachable set), several hundreds or even thousands of trajectories must be integrated. This makes the whole process slow compared to algorithms for ODEs.

2. DI right-hand side and solution representation. In the ODE models, the right-hand side of the equations is a point, and the resulting trajectory can be given as a sequence of points. In the case of a DI we must store sets instead of points. The final solution is a sequence of sets (one for each integration time step) and the required representation of the reachable set can (and should) be given as the surface of its boundary. In multidimensional case this is not a trivial problem.

3. Extension of DIs to partial differential equations is possible. However, this means that the state space is no longer Euclidean and multidimensional, but rather an abstract one like Banach, Hilbert, or Sobolev spaces. In this case, serious computational difficulties may arise because the operators involved in the model need not be bounded.

The use of multiprocessing considerably accelerates the procedure. However, the computing time may vary from one model to another. To get the contour of the set for a second-order model we need less than 1 minute. For a non-linear model of order four, the necessary CPU time may be up to 30 minutes or more.

References

[1] A. Bressan, The most likely path of a differential inclusion, Journal of Differential Equations 88 (1990) 155–174, https://doi.org/10.1016/0022-0396(90)90113-4.

[2] A. Bressan, Solutions of lower semicontinuous differential inclusions on closed sets, Rendiconti Del Seminario Matematico Dell'Università Di Padova 69 (1983) 99–107.

[3] O. Carja, Qualitative properties of the solution set of differential inclusions, Report 244 from 05/10/2011. Universidatea Alexandru Ioan Cuza, 2011, http://www.math.uaic.ro/~ocarja/idei/PN_II_ID_PCE_2011_3_0154/index.php?id=inf.

[4] P. Collins, D.S. Graca, Effective computability of solutions of differential inclusions the ten thousand monkeys approach, Journal of Universal Computer Science 15 (6) (2009) 1162–1185, https://doi.org/10.3217/jucs-015-06-1162.

[5] A.F. Filippov, Classical solutions of differential equations with multivalued right hand, SIAM Journal on Control 5 (1967) 609–621.

[6] E.B. Lee, L. Markus, Foundations of Optimal Control Theory, Wiley, New York, ISBN 978-0898748079, 1967.

[7] A. Marchaud, Sur les champs de demi-cones et les équations differielles du premier ordre, Bulletin de la Société Mathématique de France 62 (1934).

[8] A. Plis, Remark on measurable set-valued functions, Bulletin de L'Académie Polonaise Des Sciences. Série Des Sciences Mathématiques, Astronomiques Et Physiques 9 (12) (1961).

[9] E. Polak, Computational Methods in Optimization, Academic Press, New York, ISBN 0125593503, 1971.

[10] L.S. Pontryagin, The Mathematical Theory of Optimal Processes, Wiley Interscience, New York, 1962.

[11] S. Raczynski, Differential inclusion solver, in: Conference Paper: International Conference on Grand Challenges for Modeling and Simulation, SCS, San Antonio TX, 2002, 2002.

[12] Y. Takeuchi, Global Dynamical Properties of Lotka-Volterra Systems, World Scientific, 1996.

[13] A. Turowicz, Sur les trajectoires et quasitrajectoires des systemes de commande non-lineaires, Bulletin de L'Académie Polonaise Des Sciences. Série Des Sciences Mathématiques, Astronomiques Et Physiques 10 (10) (1962).

[14] A. Turowicz, Sur les zones d'emision des trajectoires et des quasitrajectoires des systemes de commande nonlineaires, Bulletin de L'Académie Polonaise Des Sciences. Série Des Sciences Mathématiques, Astronomiques Et Physiques 11 (2) (1963).

[15] T. Wazewski, On an optimal control problem differential equations and their applications, in: Conference Paper: Proceedings of the Conference Held in Prague, Publishing House of the Czechoslovak Academy of Sciences, Prague, 1963.

[16] T. Wazewski, Sur les systemes de commande non lineaires dont le contredomaine de commande n'est pas forcement convexe, Bulletin de L'Académie Polonaise Des Sciences. Série Des Sciences Mathématiques, Astronomiques Et Physiques 10 (1) (1962).

[17] T. Wazewski, Sur une genralisation de la notion des solutions d'une equation au contingent, Bulletin de L'Académie Polonaise Des Sciences. Série Des Sciences Mathématiques, Astronomiques Et Physiques 10 (1) (1962).

[18] S.K. Zaremba, Sur les équations au paratingent, Bulletin Des Sciences Mathématiques 60 (1936).

CHAPTER THREE

Market optimization and uncertainty

Abstract

This chapter deals with optimal control and uncertainty. The part that refers to optimal control is included because the methods of optimal control theory are closely connected to differential inclusions, in particular to the problem of finding reachable sets. An application in marketing is presented.

Dynamic market optimization with respect to price, advertisement, and investment is considered. The model is non-linear. Its main parameters are the elasticities with respect to price, advertisement, and consumer income. Dynamic elements have been added to the static model. Parameters like the seasonal index and consumer income are functions of time, and the whole market grows according to the investment. The tools of optimal control theory are applied to calculate optimal policies for product price, advertisement, and investment, which are controlled simultaneously. The total revenue is maximized.

The uncertainty of market parameters is discussed, and the reachable sets that correspond to parameter uncertainty are shown.

Keywords

Marketing, Uncertainty, Differential inclusion

3.1. Some remarks on optimal control theory

The problem of finding reachable sets for differential inclusions (DIs) is closely related to the problem of optimal control of dynamic systems. Here, we will discuss some aspects of the maximum principle of Pontryagin et al. [27] and give an example of, perhaps not so typical, application in marketing.

Recall the statement of the basic optimal control problem. Consider a control system described by the following (vectorial) equations:

$$\frac{dx}{dt} = f(x(t), u(t), t), \qquad (3.1)$$

where $x \in R^n$, $u \in C(x(t), t)$ $\forall t \in [0, T]$, $T > 0$. Denote $x_0 = x(0)$.

Suppose that f satisfies the Lipschitz condition and has continuous partial derivatives with respect to the state vector $x = (x_1, x_2, x_3, ..., x_n)$ and to

Reachable Sets of Dynamic Systems
https://doi.org/10.1016/B978-0-44-313384-8.00020-8

53

the control vector $u = (u_1, u_2, u_3, ..., u_m)$. $C(x, t)$ is the set of control restrictions, t represents time, and T is the final time of the control process. With such assumptions, for any integrable function $u(t) \in C(x(t), t)$ there exists a function x that satisfies (3.1), called *admissible trajectory*. The problem is to find a control function $u(t) \in C(x, t) \, \forall t \in [0, T]$ that minimizes the given object function

$$J = \psi(x(T)) + \int_0^T \phi(x(t), u(t), t) dt, \qquad (3.2)$$

where ϕ is a continuous function with continuous derivatives with respect to u and x. In the case considered in this chapter, the component ψ is not needed ($\psi \equiv 0$). In a more general case, the derivative of x may also appear as an argument of F. The control u is restricted as follows:

$$u(t) \in C(x(t), t) \, \forall t \in [0, T].$$

Here, we restrict our attention to the case without equality constraints. The conjugated vector $p = (p_1, p_2, ..., p_n)$ is introduced. By definition, it obeys the following equation:

$$\frac{dp_i}{dt} = -\sum_{j=0}^{n} \frac{\partial f_j}{\partial x_i} p_i - \frac{\partial \phi(x(t), t)}{\partial x_i}. \qquad (3.3)$$

The *Hamiltonian function* for this problem is defined in the following form:

$$H \equiv p^T f(x, u, t) + \phi(x, u, t), \qquad (3.4)$$

where p^T is the transpose of p. In the classical definition of the maximum principle, the problem is to minimize the optimization criterion. According to Pontryagin's maximum principle [27], the optimal control which minimizes the cost functional J over the time interval $[0, T]$ must maximize H for almost all $t \in [0, T]$. Thus, the problem of minimizing the object function can be replaced by the problem of finding the maximum of H with respect to u, with the restriction $u \in C(x(t), t)$, for almost all $t \in [0, T]$. The maximum principle gives the *necessary condition* of optimality.

The last problem is easier than the original one. Let us give a heuristic, rather intuitive explanation. Let us discretize the problem in time, considering k discrete time steps, with $k = 1000$. The control variable dimensionality $m = 3$. The original problem is to minimize J, which is a

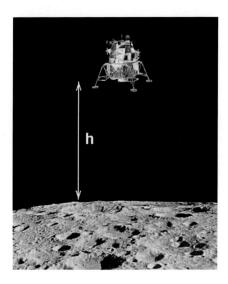

Figure 3.1 Landing on the Moon.

function of $k \times m = 3000$ variables. Using the maximum principle, we decompose the original task into k separate (algebraic) tasks of maximizing a function (Hamiltonian) of only three variables. Thus, we repeat a simple optimization procedure k times, but never solve any optimization problem of 3000 variables.

In some cases, the optimal control can be found analytically. If this is impossible, then an iterative algorithm may be used, as described later on. Let us see a simple example of an analytic solution to an optimal control problem.

3.2. Example: landing on the Moon

Suppose that the landing module is suspended over the Moon's surface, at $h = 500$ (Fig. 3.1), with no movement. It has two rocket engines that can provide a thrust P, accelerating downwards or upwards. The total thrust P is limited $(-P_m \leq P \leq P_m)$.

The problem is to land on the surface in minimal time. To avoid a crash, the final velocity must be equal to zero. The fuel consumption is also taken into account in the optimality criteria. Only vertical movement is considered. During landing, the module position h obeys the following equation:

$$\frac{d^2h}{dt^2} = \frac{P(t)}{M} - g, \tag{3.5}$$

Here, the positive acceleration is oriented upwards, M is the mass of the module (which is supposed to be constant), and g is the Moon's gravity acceleration, equal to 1.625 m/s^2. We introduce the following state and control variables: $x_1 = h$, $x_2 = h'$, and $u = P$. Thus, the movement equations are as follows (the prime mark means time differentiation):

$$\begin{cases} x_1'(t) = f_1(t) = x_2(t), \\ x_2'(t) = f_2(t) = u(t)/M - g, \end{cases} \tag{3.6}$$

with $x_1(0) = h$, $x_2(0) = 0$. Suppose that the fuel consumption is equal to $k|P(t)|$, where k is a constant. Observe that the final landing time is equal to the following integral:

$$T = \int_0^T 1\,dt. \tag{3.7}$$

Thus, the cost function to minimize is as follows:

$$J = \int_0^T 1 + k|u(t)|\,dt. \tag{3.8}$$

We have

$$\frac{\partial f_1}{\partial x_1} = 0, \ \frac{\partial f_1}{\partial x_2} = 1, \ \frac{\partial f_2}{\partial x_1} = 0, \ \frac{\partial f_2}{\partial x_2} = 0, \ \frac{\partial f_1}{\partial u} = 0, \ \frac{\partial f_2}{\partial u} = 1/M. \tag{3.9}$$

According to (3.3), the equations for the conjugated vector are

$$p_1' = 0, \ p_2' = -p_1. \tag{3.10}$$

This means that p_1 is constant, and the plot of $p_2(t)$ is a line $p_2(t) = a - tp_1$, with a being a constant.

Consequently, the Hamiltonian is as follows:

$$H = p_1 x_2 - tp_1 \left(\frac{u(t)}{M} - g \right) + 1 + k|u_1(t)|, \tag{3.11}$$

with p_1 and k constant.

First, suppose that $k = 0$ (we do not care about the fuel consumption). Recall that we must maximize H with respect to u over the interval $[0, T]$.

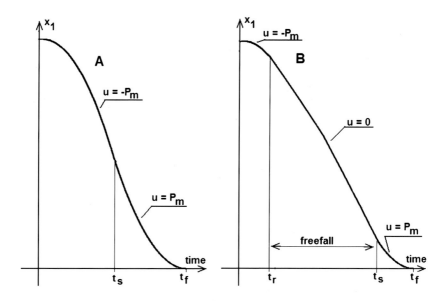

Figure 3.2 Landing on the Moon.

Observe that if p_1 is negative, then u must be equal to P_m; otherwise, it must be equal to $-P_m$. In our case p_2 can change sign only once, so we get "bang-bang" type control with only one switch point or no switching at all. In fact, this is the only conclusion we get from the maximum principle, but it is an important one. As explained below, for $k > 0$ we have two switching points, and the control is equal to zero within some time interval.

For any constant u the model trajectory is a parabola (Eq. (3.5)). Let $k = 0$. If there were no switch, then we would have only one parabola and the module could go only upwards or downwards and crash with great final velocity. So, we must have one control switch. This means that the whole trajectory consists of two parabolic segments, as shown in Fig. 3.2. Note also that, as a consequence of (3.5), $x_2(t)$ is a continuous function, so $x_1(t)$ must have a continuous derivative.

The solution for x_1 is

$$
\begin{cases}
x_1(t) = h_0 - \left(\dfrac{P_m}{M} + g\right)\dfrac{t^2}{2} & \text{for } 0 \le t_s, \\[2mm]
x_1(t) = \dfrac{1}{2}\left(\dfrac{P_m}{M} - g\right)(t_f - t)^2 & \text{for } t_s < t \le t_f,
\end{cases}
\tag{3.12}
$$

where h_0 is the initial position, t_s is the switch time, and t_f is the final time of the movement. As x_1 is continuously differentiable, we obtain the following equations:

$$\begin{cases} h_0 - \left(\dfrac{P_m}{M} + g\right)\dfrac{t_s^2}{2} = \dfrac{1}{2}\left(\dfrac{P_m}{M} - g\right)(t_f - t_s), & (a) \\[3mm] \left(\dfrac{P_m}{M} + g\right)t_s = \left(\dfrac{P_m}{M}g\right)(t_f - t_s). & (b) \end{cases} \tag{3.13}$$

Eq. (3.13a) results from the continuity of x_1 at $t = t_s$, and Eq. (3.13b) holds because $x_1(t)$ is continuously differentiable.

From the above two equations, we can calculate t_s and t_f. From (3.13b) we obtain

$$t_f = \frac{2P_m}{P - Mg}t_s. \tag{3.14}$$

Substituting this into (3.13a) and rearranging, we get

$$t_s = \sqrt{\frac{h_0 M}{P + Mg}}. \tag{3.15}$$

Now, we incorporate the fuel consumption into the cost function, i.e., $k > 0$. Recall that we must maximize the Hamiltonian with respect to u. Now, the control u appears also in the last term of (3.11). Remember that p_2 intersects the horizontal axis (zero value). When p_2 is sufficiently small, the term $k|u|$ dominates, so the maximum of the Hamiltonian is reached for $u = 0$ because this maximizes the negative absolute value of u. So, the trajectory is composed of three stages, $u = -P_m$, $u = 0$, and $u = P_m$ (see Fig. 3.2B). From the continuity assumptions we can calculate, in a similar way, the switching times t_r and t_s and the final time t_f. This will require somewhat more complicated geometrical considerations, but this is a rather geometric and algebraic problem.

In the more complicated, multidimensional case, the (numerical) problem is that we do not have the initial conditions for vector p and we cannot integrate (3.3) forward in time. Instead, we have the final conditions for p. These conditions can be derived from the *transversality conditions* (Lee and Markus [16]). This leads to the *two-point boundary value problem*. Anyway, the control u is still unknown, so we cannot integrate the state and conjugated equations. A possible solution consists in assuming an arbitrary admissible control, integrating the state equations, and then looking for a way to improve the control. The process is repeated with the new control function.

To be able to improve the control, we must know the direction in which the control should change. This direction is given by the gradient of the Hamiltonian.

The gradient of H is given by the following expression:

$$grad\,H = p \cdot g, \quad \text{where } g = \left(\frac{\partial f_1}{\partial u_1}, \frac{\partial f_2}{\partial u_2}, ..., \frac{\partial f_n}{\partial u_n} \right) \qquad (3.16)$$

for each time step.

In other words, $grad\,H$ is, at the same time, the search direction in the control space. To calculate $grad\,H$ for all $t \in [0, T]$, a trajectory $x(t)$ must be calculated and stored, starting with a given initial condition for x and an arbitrary control. The conjugated vector equations are calculated backwards, starting with the final conditions for p. The problem is that the control function is not defined yet, and we do not have the initial conditions for vector p. Instead, we have the final conditions for p.

According to the *transversality conditions* for the problem (fixed time, free end point), the final condition for the conjugated vector is, in our case, $p_1(T) = 1, p_2(T) = 0, ..., p_n(T) = 0$. A possible iterative algorithm is as follows (Polak [26]).

1. *Select an arbitrary control function u(t) (it should be an approximation of an optimal control, if we can get one), integrate Eq. (3.1) over [0, T], and store the state for each time step.*

2. *Starting with the final condition for p, calculate and store the trajectory of p(t) backwards in time, using Eq. (3.3). Simultaneously, calculate and store grad H.*

3. *Once grad H has been defined, adjust control u in the search direction. Now, having this new control function, go to step 1.*

The stop condition for this algorithm can be that a small enough value of $grad\,H$ is obtained or that the object function does not improve any further.

This is the simplest possible version of the optimization algorithm. To accelerate the search, we can replace u in step 3 by the new value which maximizes the Hamiltonian, instead of advancing a small step in the search direction. This, however, may provoke stability problems for non-linear models. Another common modification is to implement the method of *steepest descent* with conjugated gradients in the control space (Polak [26]). In our case a simple steepest descent method was implemented.

3.3. Simulation and optimization

The market is a complex, non–linear, socioeconomic system. Models of such objects are known as *soft system models*. Looking at the annals of the huge literature in the field and comparing the models, one can observe that the market models are almost always completely different from each other. Many publications in the field of marketing, like those of Lilien and Kotler [18], offer a large number of models (linear, non–linear, deterministic, stochastic, static, dynamic, etc.). Any model should be related to the corresponding experimental frame (Zeigler [34]). To be realistic (we do not use the adjective "valid"), the experimental frame must include the most important marketing parameters. The model must be clear enough to be understandable for marketing and management staff, and it must be useful while carrying out simulation experiments. If we also want to find the optimal control of the market, then the simulations must be embedded in optimization algorithms and must run fast enough. Consult also the article [23].

Finally, another requirement is the ability to simulate the market dynamics. This means that the optimization must be dynamic, like in optimal control problems. In this chapter, the non–linear model with elasticities with respect to *price, advertisement*, and *consumer income* has been modified with dynamic inertia added to the market response. The resulting model is subject to multivariable dynamic optimization. Other important variables are included, like the *seasonal index, overall market growth*, and *investment*, being functions of time.

The search for market models for simulation purposes dates from the early 1960s, while significant publications began to appear in the late 1960s. King [13] gives a comparison of *iconic, analog*, and *symbolic models*. Recall that an iconic model represents reality on a smaller scale, an analogical model shows reality in maps and diagrams, and a symbolic model uses mathematical expressions to portray reality. Montgomery and Urban [19] discuss the *descriptive, predictive*, and *normative* models, where by *descriptive* we mean models that consist largely of diagrams and maps or charts designed to describe a real-world system. See also Stanovich [32]. Predictive models are used in predictive analysis to create a statistical model of future behavior, and normative models evaluate alternative solutions in order to answer the question "What is going on?" and to suggest what ought to be done or how things should work.

We should distinguish between *macromarketing* and *micromarketing* models (Lilien and Kotler [18]). Macromarketing addresses big and important issues at the nexus of marketing and society, while micromarketing refers to marketing strategies which are customized to local markets, to different market segments, or to the individual customer (Shapiro et al. [28]). Here, we deal with a symbolic micromarketing model that can be implemented in computer simulations and then used in optimal control algorithms.

Optimal control in marketing is not a new topic. There are many publications in the field, like, for example, Bertsimas and Lo [3]. In that article we can find the application of Bellman's dynamic programming approach to the problem of the price impact on the dynamic trading on the stock exchange. A tutorial and survey on the relevant technical literature on models of economic growth can be found in Burmeister and Dobell [5]. Feichtinger et al. [9] consider a similar problem for a general market and optimal advertisement policy. In that article a detailed formulation of an implementation of the maximum principle of Pontryagin is discussed. Yuanguo [33] uses Bellman's principle of optimality, and derives the principle of optimality for fuzzy optimal control. This is applied to a portfolio selection model.

The general approach to optimization of economic systems including households' consumption, labor supply, production, and government policies can be found in the book of Dixit [8]. As the author explains, the methodology is based on the use of verbal and geometric arguments, but with an eye toward mathematical sharpening and generalization. Concepts of optimization with respect to variables such as price, consumer income, and quantities of goods can make this methodology useful in marketing problems.

Chow [6] in his book describes the application of the *Lagrange method* to the optimization of economic systems in general. Instead of using dynamic programming, the author chooses the method of Lagrange multipliers in the optimization task. A number of topics in economics, including economic growth, macroeconomics, microeconomics, finance, and dynamic games, are treated. The book also provides examples, starting with simple problems and then moving to general propositions.

The Lagrange method is a powerful optimization tool, but needs some additional conditions and modifications to be applied in practice. Namely, optimization of models with constraints should be treated rather with the tools of optimal control theory, like the maximum principle used in this chapter. The maximum principle was derived from the Lagrange method and the calculus of variations, and applied to dynamic optimization problems with constraints.

Konno and Yamazaki [14] present a large-scale optimization problem of a stock market with more than 1000 stocks, and show that the problem can be solved using the *absolute deviation risk*, called *L1 risk function*. In fact, the stock market has its own properties that need methods oriented to that particular problem.

Another approach to stock market optimization can be found in Speranza [31]. The paper describes an application of an optimization algorithm to the Milan stock market, taking into account portfolios with transaction costs, minimum transaction units, and limits on minimum holdings. The author points out that the presence of integer variables dramatically increases the computational complexity.

Karatzas et al. [12] consider a general consumption/investment problem for an agent whose actions cannot affect the market prices, and who strives to maximize the total expected discounted utility of the consumption as well as terminal wealth. They decompose the problem by separately maximizing the utility of consumption only and maximizing the utility of terminal wealth, and then appropriately recombine these sub-problems. Such decomposition may work in some special cases. However, note that in a general case it may fail. Observe that in the present chapter we do not assume any possibility of decomposing the optimization problem into any partial sub-problems. In another paper of Karatzas [11] we can find a unified approach, based on stochastic analysis, to the problems of pricing, consumption/investment, and equilibrium in a financial market with asset prices modeled by continuous semi-martingales, and a similar problem decomposition. The Hamilton–Jacobi–Bellman equation of dynamic programming associated with this problem is reduced to the study of two linear equations. The results of this analysis lead to an explicit computation of the portfolio that maximizes the capital growth rate from investment and to a precise expression for the maximal growth rate.

Korn and Korn [15] in their book offer a collection of graduate studies in mathematics, including the mean-variance approach in a one-period model, a continuous-time market model, pricing of exotic options, and numerical algorithms. The marketing models in the book are shown from a mathematical point of view and reflect the state of the art in the field (as for year 2001).

Gomes Salema et al. [10] contemplate generic reverse logistics and distribution networks where capacity limits, multiproduct management, and uncertainty on product demand and returns are considered. A mixed integer formulation is developed, using standard business-to-business (B2B) techniques. The model is applied to an illustrative case. To learn more on B2B strategies, consult, for example, Morris et al. [21].

There are many other publications on optimal market control, most of them applied to the stock market or to markets of specific goods. It seems that the dominant tools are dynamic programming and the Lagrange method.

3.4. The model

The market model itself is not the main topic of the present chapter (except perhaps some dynamics added). Let us start with the model taken from the book of Lilien and Kotler [18] mentioned earlier. We discuss an application of the maximum principle to the problem of dynamic optimization of a non-linear revenue model with respect to price, advertisement, and investment. The elasticity-based model has been modified to reflect the market dynamics, including the inertia with respect to the control variables. This is a non-linear model of the demand and revenue, with the following parameters: *price, advertisement, seasonal index, overall market growth, consumer income, production or acquisition cost,* and the *market elasticities with respect to the price, advertisement, and consumer income.* The demand model is as follows:

$$q(p, a, y, v, g, t) = q_0 v(t) s(t) \left(\frac{p(t)}{p(0)} \right)^{e_p} \left(\frac{a(t)}{a(0)v(t)s(t)} \right)^{e_a} \left(\frac{v(t)}{v(0)} \right)^{e_y}, \quad (3.17)$$

where q is the demand, q_0 is the initial or reference demand, t is the time, $s(t)$ is the overall market size, $p(t)$ is the price, $a(t)$ is the advertisement per time unit, $y(t)$ is the consumer income, $v(t)$ is the seasonal index, and e_p, e_a, and e_y are the market elasticities with respect to price, advertisement, and consumer income, respectively.

In the original model of Lilien and Kotler, the market size $s(t)$ is supposed to be equal to $(1 + g)t$, where g is the market growth rate. We use a more general function $s(t)$ instead, to be able to link the market growth to the investment that the company may make in order to expand (new installations, infrastructure, etc.).

Normally, the price elasticity is negative (price increase means less demand), and the elasticities for advertisement and consumer income are positive. Note that the term $v(t)s(t)$ multiplies the demand, and appears also in the denominator of the advertisement impact term. This means that if the market and the seasonal index grow, then we must spend more on

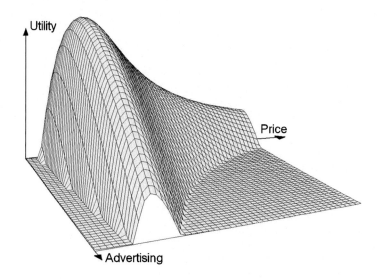

Figure 3.3 The revenue as function of price and advertising.

advertisement to achieve the same effect. We do not consider any storage or warehouse mechanism, so it is supposed that the sales are equal to the demand.

The model includes the competition in some indirect way. Observe that the market elasticities with respect to price and consumer income are influenced by the competition. If, for example, a company provides water to a village and has no competitors, then the elasticity e_p will approach zero because the consumers must consume a certain amount of water anyway. However, if there is a competitor in the region, then this elasticity must be negative because the price will influence the demand for water provided by the company. Another way to introduce competitors to the model would be to simulate the performance of two or more competing companies. However, if we intend to optimize the market for one company, then we must simulate the optimal strategy of the competitors as well. This leads to a differential game and not just a single optimization problem.

The revenue can be calculated as follows (see Fig. 3.3):

$$U = (1 - b)e - a, \quad V = e - a, \quad e = (p - c)q. \tag{3.18}$$

Variable U will be called total revenue, and V will be called revenue. Here, c is the unit cost of the product (production, acquisition cost), a is the advertisement, b is the investment factor, and q is given by formula (3.16).

The investment factor b $(0 \leq b \leq 1)$ is supposed to be part of the utility e. The factor b defines the market growth rate, and is proportional to the product be. It will be one of our control variables. In our equations, the market size is relative with initial value equal to 1. This means that if the company does not invest in the market, then the market size has no influence on the model trajectory (remains constant). The final total revenue is the sum (integral) of U over the time interval under consideration. Fig. 3.3 shows the shape of the profit (utility) e as a function of the price and advertisement, while the time and other variables are fixed. It can be seen that the utility function has a maximum that determines the optimal price and advertisement at the moment. This is the static case.

Now, consider a fixed time interval $[0, T]$. To calculate the total revenue we must integrate U (see (3.17)) over $[0, T]$. We should also take into account the market inertia with respect to the price and advertisement. In our model we added first-order inertia with time constants T_p and T_a, for the price and advertisement variables, respectively. So, the model becomes of the fourth order, described by the following equations:

$$\begin{cases} v = p(t) - c(t))q(z, w, v, s, t) - a(t), & \\ r' = (1 - b(t)v, & \text{total revenue} \\ w' = (a(t) - w(t))/T_a, & \text{advertisement inertia} \\ z' = (p(t) - z(t)/T_p, & \text{price impact inertia} \\ s' = kb(t)(p(t) - c(t))q(z, w, v, s, t), & \text{market size} \end{cases} \quad (3.19)$$

where r is the total revenue, w is the advertisement inertia, also called *consumer goodwill* related to advertisement, and z is the price impact with inertia. The coefficient k tells how fast the market grows due to the investment; its dimension is 1/(currency unit [CU]). For example, $k = 0.001$ means that one invested CU makes the market grow by 0.001 (relative) in one time unit.

The investment in the market growth cannot be negative, so the control b in (3.17) is set equal to zero when $(p - c)q$ becomes negative. It is supposed that $0 \leq b \leq 1$.

Note that in (3.18) the demand depends on z and w instead of the price p and advertisement a, taken at the current time instant. The object functional we want to maximize is the final revenue $r(T)$, and the optimization is carried out with respect to the price p, advertisement a (through the corresponding controls, defined later), and the investment factor b. This way, we obtain a system of four non-linear differential equations with three control variables. Note that the variables r and s are closely connected to each other, and also depend on w and z. The model is rather simple, but its optimization is not a trivial task. In the optimization algorithm the control variables are as follows: u_1 tells which part of the initial (reference) advertisement is used as the actual advertisement a ($a = u_1 a_0$); u_2 tells which part of the initial (reference) price is applied; and u_3 tells which part of the revenue V (see (3.17)) is used as the investment (investment factor b).

The advertisement $a(t)$ does not depend explicitly on the investment factor u_3. However, the model includes a number of restrictions that result from the market logic. Of course, the price, advertisement, and investment cannot be negative. Moreover, the advertisement cannot be greater than $U - (1 - b)e$. This makes the advertisement depend on U and on the investment factor b, through the corresponding restriction.

3.5. Computer implementation

Our optimization process is iterative, and the model is non-linear of order four with up to three controls, so the whole process is rather slow, with several minutes of computing time needed. The Runge–Kutta integration method was initially implemented, but this resulted in rather slow computations. As we need rather qualitative results, the integration of the model trajectories has been done by a fast, but less exact, Euler method, with a reasonable time step (1000 steps in the simulation time interval). Little difference was observed between the results obtained by the two methods.

We assume the same fixed time horizon equal to 365 (days) in all simulations. In addition to the model equations in (3.18), some simple arithmetic operations have been added to avoid negative revenue growth. In other words, we do not consider the possibility of negative revenue at any moment, and all the advertisement and investment must come from the instant sales income. Such restrictions, as well as other non-negativity restrictions imposed on the controls, complicate the optimization process. Recall that one of the assumptions of the maximum principle of Pontryagin is that the right-hand sides of the equations in (3.18) are continuously differentiable. In our case this is not exactly true. So, all the results should be treated as approximations of the optimal solution (anyway, the algorithm is iterative), and some of the resulting curves show certain irregularities (they are not just "nice and smooth").

The initial conditions for the state vector have been fixed to $(0, 0, 0, 1)$ for total revenue r, advertisement goodwill w, price impact with inertia z, and the relative market size s, respectively. Let us show results of four experiments, where the market is optimized with respect to price, advertisement policy, price and advertisement, and finally all three controls simultaneously. We also want to see the impact of the seasonal index and the consumer income on the optimization results. So, the seasonal index is set equal to one, except of two intervals of time: $(100, 128)$ and $(160, 167)$ days. In the first interval the index jumps to 3, and in the second it falls down to 0.5. The relative consumer income also changes; it is equal to one everywhere, except for the interval $(200, 214)$, where its value is equal to 2.5. We will see that the optimization algorithm has a "predictive" ability, namely, the corresponding control changes anticipate the changes of these two indices.

Model parameters are as follows:
Initial (reference) sales = 70,000 items
Production cost = 0.8 CU
Initial (reference) price = 1.0 CU
Initial (reference) advertisement = 10,000 CU
Market elasticity for price = -2
Market elasticity for advertisement = 0.5
Market elasticity for consumer income = 0.3
Time constant for advertisement inertia (goodwill) = 14 days
Time constant for price impact inertia = 2 days
Initial market size (relative) = 1
Final simulation time = 365 days
Market growth factor = 0.00000005 (this means that one invested CU makes the relative market size grow by 0.00000005 per time unit)

The set $C(t)$ of (3.1) is defined by the set of the restrictions imposed on the model. In particular, model variables that logically cannot be negative are restricted to non-negative values. The investment cannot be positive if the revenue V is equal to zero. Also, the advertisement cannot be greater than $U - (1 - b)e$ (Eq. (3.17)).

It should be noted that this is an optimization problem with fixed time horizon. In other words, the company starts to sell goods and disappears after a given time interval. The final simulation time is 365 days. The behavior of the control variables for the last few days appears to be somewhat strange. Obviously, the optimal control for investment and advertisement near the end of the interval fall down; there is no reason to waste money

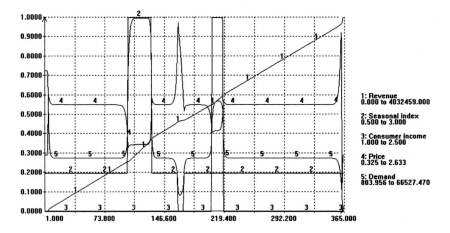

Figure 3.4 Price control only.

if the activities terminate. On the other hand, the algorithm increases the price during the last days of activity. The sales do not drop immediately because of the market inertia with respect to price, so a better final outcome is obtained. All curves are normalized to the interval [0, 1]. The real ranges are indicated in the legend to the right of the plot.

Fig. 3.4 shows the results of market optimization with respect to the price of the product. Note that when the seasonal index and the demand grow, the recommended policy is to lower the price. Anyway, the revenue (curve 1) grows faster in that period. When the seasonal index becomes low, the optimal strategy is to first increase the price and then reduce it rapidly. As for the period of increased consumer income, the optimal policy is to lower the price to make the demand higher. All these changes should be made with anticipation because of the market inertia.

The results of advertisement policy optimization are shown in Fig. 3.5. In the period of increased seasonal index, it is recommended to increase the advertisement. When the consumer income grows, the advertisement should grow as well.

If we optimize both the price and advertisement simultaneously, the curves are similar (Fig. 3.6). The total revenue at the end of the period is, in this case, greater than when price and advertisement are optimized separately. Fig. 3.7 shows the comparison between the applied optimization modes (revenue value).

Now, suppose that the company invests in the market (new marketing places, infrastructure). The income from sales can be invested or "con-

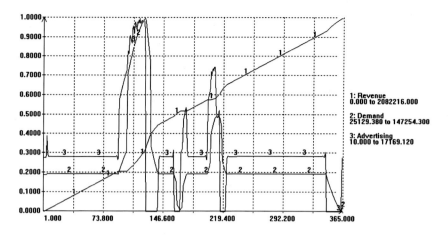

Figure 3.5 Advertisement control only.

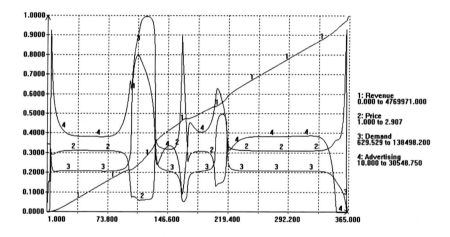

Figure 3.6 Price and advertisement control.

sumed" immediately. If we invest, the market and the future income grow. Logically, the optimal policy should be to invest at the very beginning, to get more income later on. The optimization algorithm generates a "bang-bang" type control for the investment. First, all the income is spent on investment and advertisement, and then the investment is set equal to zero. See the corresponding curves in Fig. 3.7.

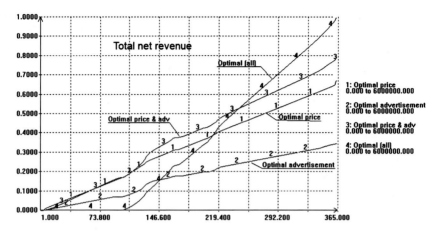

Figure 3.7 Revenue curves. Comparison between optimization with respect to price only, advertisement only, both price and advertisement, and all control variables.

3.6. Market model with uncertainty

Uncertain consumer behavior makes it difficult to predict future events and elaborate good marketing strategies. Frequently, uncertainty is modeled using stochastic variables. Our approach is quite different. The dynamic market with uncertain parameters is treated using *differential inclusions* (DIs). This permits us to determine the corresponding reachable sets. This is not a statistical analysis. We are looking for the reachable sets, which are deterministic objects. The purpose of the research is to find a way to obtain and visualize the reachable sets, in order to know the limits for the important marketing variables.

3.6.1 Uncertainty problem

Our modeling method consists in defining the *DI* related to the model and to find its reachable set. The main uncertain parameter is the share of investment, being a part of the revenue. As an additional result we can also define the optimal investment strategy. This section is focused on the treatment of uncertainty by means of the reachable sets, rather than stochastic modeling. The model of the demand is not our main topic. It is taken from the literature. However, we add some dynamic properties to the original, static model.

Applications of DIs and reachable set determination to marketing problems are not very common in the known literature. Beard [2] presents an application of DIs to a very specific problem of fertilizer carryover, based on the *Liebig law* or the *law of the minimum*. This law was developed in agricultural science by Justus von Liebig [20]. It states that growth is controlled not by the total amount of resources available, but by the scarcest resource (limiting factor).

The approach of Read et al. [25] to the uncertainty in marketing is more heuristic. Two groups of experts and managers were asked to think aloud as they made marketing decisions in exactly the same unpredictable situation. The results show significant differences in heuristics used by the two groups. While those without entrepreneurial expertise depended primarily on predictive techniques, experienced entrepreneurs tended to use an effectual or non-predictive logic to tackle uncertain market elements and to construct novel markets with committed stakeholders.

The problem of market uncertainty is also addressed by Slovik [30]. The paper describes the market uncertainty theory comprising the market uncertainty theorem and the notion of heterogeneity of market uncertainty, as well as policy recommendations relevant for rating agencies, financial institutions, and public authorities. Lien [17] considers the problem of uncertainty from a cultural and an ethnographic point of view. However, as in the papers of Read and Slovik mentioned above, the problem is stated and described in a textual way, with no models (logical or mathematical) proposed. This does not mean that their or our approach is better, but means rather that the different research tools can hardly be compared to each other.

A conceptual framework can be found in Simos et al. [29]. They propose a framework of antecedents and market performance of emergent marketing strategies. The model is tested using a sample of 214 UK enterprises. The results suggest that dimensions of market uncertainty and strategic feedback systems influence the formation of emergent marketing strategies.

Courtney et al. [7] present another approach to market uncertainty. In their article, four levels of uncertainty are considered: (1) a clear enough future, (2) alternate futures, (3) a range of futures, and (4) true ambiguity, when multiple dimensions of uncertainty interact and create an environment that is virtually impossible to predict. Compared to a more mathematically rigorous DI approach, the most similar is level 3 of uncertainty, where a range of possible future outcomes for marketing data can be known. Heuristic strategies for the different levels of uncertainty are proposed. The general concepts of the work may be useful in marketing practice, though no formal market model is used.

The paper of Brace et al. [4] is one of numerous works on stochastic market models using the Heath–Jarrow–Morton (HJM) framework. This and similar methods are used to analyze forward rates and pricing statistical behavior. It is difficult to compare the DI approach to these methods. Our approach is very different. The main point is that in marketing and in more general soft system models, not everything can be treated using statistical frameworks.

The method presented here can also be used to obtain optimal investment strategies. Anyway, to determine reachable sets we use some concepts taken from optimal control theory.

We calculate the reachable sets for all possible investment strategies to assess the impact of the uncertainty in important market variables. It should be emphasized that our approach to uncertainty has nothing to do with randomness or stochastic models. If the value of a parameter is uncertain, this means that it can take values from some interval, but this does not mean that it is a random variable. Such a parameter does not have any probability distribution. The value changes may be caused by any internal or external agents. For example, in the stock market the information about the actual share price can be obtained by means of observations and predictions. However, this also can be false information that is introduced intentionally. To obtain the reachable sets, we use the DI solver; see Chapter 2 for more details about this tool.

3.6.2 Differential inclusions

Here, we only recall some main terms. DIs have been treated in more detail and with more mathematical rigor in Chapter 1.

Consider a dynamic model given in the form of an ordinary differential equation (state equation),

$$\frac{dx}{dt} = f(x, u, t), \tag{3.20}$$

where $x \in R^n$ is the state vector, t is the time, f is a vector-valued function, and $u \in R^m$ is an external variable, called control in automatic control theory. Suppose that the value of the control u is restricted so that $u(t) \in C(x, t)$, where $C(x, t) \subset R^m$. For each fixed x and t, the function f maps the set C (all possible values of u) into a set $F \subset R^n$. In this way we obtain the following condition:

$$\frac{dx}{dt} \in F(x, t). \tag{3.21}$$

What we obtained is a *differential inclusion* (DI). Here, F is a set-valued function. The right-hand side of Eq. (3.20) defines the set F, when u scans all its possible values. For more information on DIs, consult Aubin and Cellina [1], Zaremba [35], and Chapter 1 of the present book.

In this very natural way, we see that the uncertainty in dynamic system modeling leads to DIs as a corresponding mathematical tool. Namely, if any model parameter has an uncertain value, it can be treated as a limited control that belongs to a given interval or set. This determines a DI.

3.6.3 Differential inclusion solver

The *DI solver* calculates and displays the solution to a given DI. The software was developed by the author (consult [22]). This tool is described in detail in Chapter 2. The present research can hardly be compared to other similar published works because the DI solver has never been used before in marketing models and the images of the corresponding reachable sets can hardly be found in the literature.

3.6.4 Application to a market model

In this section, we use a simplified market model, similar to that of Section 3.4.

The parameters are: *price, advertisement*, the *seasonal index, overall market growth, consumer income, production or acquisition cost*, and the *market elasticities with respect to price, advertisement, and consumer income* (Lilien and Kotler [18]). The demand model is as follows (we repeat the formula from Section 3.4):

$$q(p, a, y, v, g, t) = q_0 v(t) s(t) \left(\frac{p(t)}{p(0)} \right)^{e_p} \left(\frac{a(t)}{a(0) v(t) s(t)} \right)^{e_a} \left(\frac{v(t)}{v(0)} \right)^{e_y}, \tag{3.22}$$

where q is the demand, q_0 is the initial or reference demand, t is the time, s is the overall market size, p is the price, a is the advertisement per time unit, y is the consumer income, v is the seasonal index, and e_p, e_a, and e_y are the market elasticity with respect to price, advertisement, and consumer income, respectively.

The net revenue U is calculated as follows:

$$u = (1 - b)e - a, \quad \text{where } e = (p - c)q, \tag{3.23}$$

where c is the unit cost of the product (production, acquisition cost), a is the advertisement, b is the investment factor, p is the price to the public, and q is given by formula (3.22). The investment factor b $(0 \leq b \leq 1)$ determines what part of the utility e is being invested in the market growth. The market growth rate is proportional to the product be. This will be one of our control variables. In our equations, the market size is relative, with initial condition equal to 1. If the company does not invest in the market ($b = 0$), then the market size has no influence on the model trajectory (remains constant). The final total revenue is the sum (integral) of U over the time interval under consideration.

To make the model more realistic, we suppose that the market growth rate follows the investment with certain inertia. The basic equations are as follows.

$$
\begin{cases}
\dfrac{dr}{dt} = (1 - b)\dfrac{v}{v_0} & \text{accumulated revenue growth,} \\[1.5ex]
\text{where} \\[0.5ex]
v = (p - c)q(z, w, y, v, s, t) - a, & \text{(3.24)} \\[1.5ex]
\dfrac{ds}{dt} = (x - s)/T_s & \text{market size growth,} \\[1.5ex]
\dfrac{dx}{dt} = gbv/v_0 & \text{accumulated investment.}
\end{cases}
$$

Here, r is the relative accumulated net revenue, s is the relative market size, x is an auxiliary variable, T_c is the time constant of the market growth inertia, and g is a constant that defines the impact of the investment on the market expansion, namely, g tells us how big the relative market growth is per CU invested. All other variables may change in time. Observe that the additional state variable x has been introduced to manage the inertia.

The final accumulated revenue and the market size strongly depend on the changes of the investment control variable b. Thus, it is important to be able to see the limits of the revenue and the changes in market size due to the investment strategy. Note that if $b \equiv 1$ (maximal investment), we have no revenue available, and if $b = 0$, the market does not grow. This also reduces the available profit.

We calculate the reachable sets for all possible investment strategies to assess the impact of the uncertainty on important market variables. As stated before, our approach to uncertainty has nothing to do with randomness or stochastic models. If the value of a parameter is uncertain, this means that it can take values from some interval, but this does not mean that it is a random variable. Such a parameter may have no probability distribution. The value changes may be caused by any internal or external agents. To obtain the reachable sets, we use the DI solver; see Chapter 2 for more details about this tool.

Model (3.24) has the form of a control system. This defines the corresponding DI. In the investment strategy problem, the control variable is $b(t)$, the share of investment applied out of the total revenue (Experiment 1 below). The uncertainty of the other model variables can be treated in a similar way, letting the variable change within a given interval. Below we present the results of DI applications in problems with investment uncertainty. For more information, see also [24].

3.6.5 Experiment 1: investment

Our aim is to obtain the reachable set in the revenue-market plane for all possible strategies of investment, with final time fixed. As an additional result, the DI solver provides the control function (investment as a function of time) for any point on the boundary of the reachable set. Thus, we can obtain the optimal strategy that maximizes the total revenue or the final market size.

The model is given by the equations in (3.24), where b is the control variable that can change in the range of 0 to 1.

The model parameters are as follows:

Initial sales = 70,000 items
Initial (reference) advertisement = 10,000 per time unit (a day)
Product cost = 0.8 CU
Initial relative market size = 1
Unit price = 1
Unit initial relative price = 1.2
Time constant for investment inertia $T_c = 30$ days
$g = 0.006$
Market elasticity for the product price = -2
Market elasticity for advertisement impact = 0.6
Final simulation time = 365 days

Fig. 3.8 shows the image of the reachable set for the revenue and market size at the final simulation time, with the investment control b included in [0, 1]. It should be noted that in our case the dimensionality of the state space is equal to three. Thus, to show the complete reachable set we

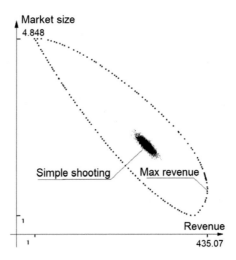

Figure 3.8 Reachable set for Experiment 1.

should display it in the 4D time-state space, which is quite difficult. If we limit the displayed state space components to market size-revenue, then the boundary of the reachable set is not always seen as just one contour because for each time instant the 2D image is a projection of a 3D reachable region. Fortunately, in this case the boundary is clearly seen, perhaps because the control vector dimensionality is equal to 1. In the same figure we can also see the set of reachable points obtained by simple (primitive) random shooting (see Chapter 2) (10,000 trajectories integrated). It is clear that the simple shooting provides a wrong assessment of the reachable set.

Fig. 3.8 also shows the advantage of the DI approach compared to other methods that use models with random variables to treat uncertainty (stochastic approach, random scan of the reachable set interior). Whatever the distribution of the uncertain parameter would be, the estimate of the reachable set with such simple random scan will be wrong. Moreover, recall that uncertain parameters are not necessarily random. An uncertain parameter may have no probability distribution at all, and may have a quite deterministic value.

Fig. 3.9 depicts a 3D image of the reachable set in the time-state space. Once the set has been displayed, the user can select any point of its boundary and see the corresponding control. Selecting the point with maximal revenue we obtain a control of "bang-bang" type, equal to one for time between 0 and 97 days and equal to zero between 97 and 365 days.

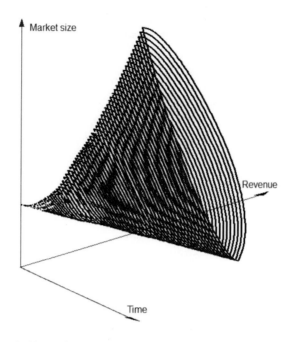

Figure 3.9 Reachable set for Experiment 1.

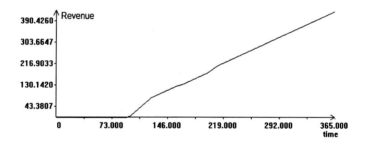

Figure 3.10 Net revenue accumulated with optimal investment strategy (relative).

The changes in the corresponding accumulated net revenue are shown in Fig. 3.10.

3.6.6 Experiment 2: uncertain price elasticity

In this experiment we suppose that the value of the market elasticity with respect to price (e_p) is uncertain and may change by ±20% of its original value. Using model parameters as in Experiment 1, with variable investment control in [0, 1] and the uncertain (variable) e_p, we obtain a larger

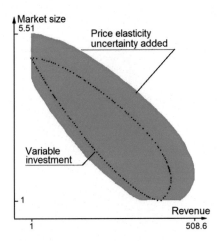

Figure 3.11 Reachable set with variable investment and uncertain price elasticity.

reachable set. In this case, the set is more complicated (2D projection of a 3D region). This provides a cloud of reachable points and not in a simple contour. Post-processing of this cloud gives us an assessment of the reachable set, shown as the gray region in Fig. 3.11. To compare with Experiment 1, the boundary points of the previous reachable set are also shown in the same figure. Observe that both the net revenue and the market size may reach greater values than those obtained with fixed elasticity e_p.

In general, it can be seen that our market needs the optimized investment strategy to reach maximal revenue. Note that we have a fixed final simulation time, as if the company sells during the given time interval and then disappears. Experiments with moving time horizon can be done, but may be more complicated and time consuming.

3.6.7 Experiment 3: model sensitivity

In our experiment, the influence of the parameters is being treated dynamically, i.e., we allow the parameters to be functions of time, changing along each model trajectory. The reachable sets obtained in this way show the influence of the uncertain parameters, similar to the sensitivity analysis. Recall that the basic sensitivity analysis shows the possible changes in the model state due to the changes of the initial conditions and the (constant) model parameters. Our problem statement and that of sensitivity analysis methods are quite different from each other, so it is difficult to compare these methods. We do not pretend to prove that the DI applications are better or worse based on the sensitivity analysis results. They are just different. Here, we use functional sensitivity, see Chapter 1.

Sensitivity analysis is not the main topic of this chapter, so we will not provide any review of the (huge) body of literature in the field. For a comprehensive review, see Chapter 1 of this book and Hamby (1994). A common way to conduct sensitivity analysis is to assign probability density functions to each model parameter and assess the resulting outcome variance with respect to the corresponding parameters. Though the advanced methods in sensitivity research are quite sophisticated and perhaps more accurate, we will use in this comparison a basic sensitivity concept. One of the most well-known methods is the general error propagation formula, i.e.,

$$V(Y) = \sum_{i=1}^{n} \left(\frac{\partial Y}{\partial X_i} \right)^2 V(X_i), \tag{3.25}$$

where Y is the model output (the outcome being analyzed), $V(Y)$ is the variance of Y, the model parameters are $X_1, \ldots, X_n$, and $V(X_i)$ is the variance of the parameter X_i. In our case, Y may be the value of any of the model state variables at the end point of the simulated trajectory. We do not have any algebraic expression for Y, whose value is the result of model simulation. However, as the calculation of a single trajectory takes only a fraction of a second, the partial derivatives can be easily calculated by numerical differentiation. Let us do this for the values of the final revenue and market size, with respect to investment and the price elasticity parameter, with the model parameters as in Experiment 2.

The investment control variable is supposed to belong to [0, 1], and the price elasticity will change by ±20%, belonging to the interval [−2.4, −1.6]. Assuming a uniform probability distribution, the variance is equal to 0.0833 and 0.0533 for the investment and price elasticity, respectively. The corresponding standard deviations are 0.288 and 0.23, respectively. The calculations are performed for the model trajectory obtained with the midpoint values for both parameters. The results are shown in Fig. 3.12 (functional sensitivity). The bold rectangle shows the region of plus/minus one standard deviation for the outcome. It can be seen that this region is little informative, and obviously very different from the reachable set (gray area).

Formula (3.25) has been obtained using model linearization. As our model is non-linear, a better, direct way to see the influence of the parameters is simply integrating model equations and changing the parameters,

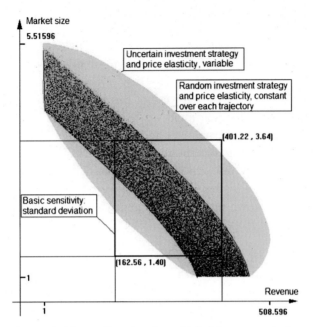

Figure 3.12 The reachable set (functional sensitivity) for uncertain investment and price elasticity and conventional sensitivity analysis results.

each of them being constant over the trajectory. The result is shown in Fig. 3.12 as a dotted area. Observe that the true reachable set is greater than the region obtained in this way. For more complex models, in particular for oscillating systems, this difference may be greater. The determination of the reachable set can be treated as the result of the functional sensitivity analysis, defined in Chapter 1.

3.7. Conclusion

Control theory provides useful tools for dynamic optimization of models in many fields of research. Once we have the model equations, we can try to apply an optimization algorithm to them. The problem is that in the real world, not everything can be modeled with differential equations. In marketing, optimization methods work to some extent, due to the intense research that has been done in the past decades. This research resulted in a variety of models that can be used in dynamic simulation and optimization.

Though the presented application of optimal control theory does not use DIs, it has been included in this book because we use some mechanism of the maximum principle in reachable set calculations. Anyway, each trajectory that scans the boundary of a reachable set is optimal in some sense. Recall that if a point in the model state space is accessible from the initial point by a trajectory of a dynamical system (with appropriate regularity assumptions), then it is also accessible in optimal (minimal) time.

The market model presented here is rather simple, but finding its optimal policies is not a trivial task. The software is rather in the experimental stage. It permits simultaneous dynamic optimization with respect to three principal marketing parameters. The model itself was not the main topic of this chapter. However, its modification (price and advertisement response inertia) represents perhaps a relevant contribution. The main topic is the application of the control theory methodology (the maximum principle) to the multivariable dynamic market optimization. The point we wish to make is that due to the restrictions imposed on control variables, the Pontryagin maximum principle is the most adequate tool. The algorithm and the experimental version of the software may form a core of a more practical and commercial marketing optimization package.

The optimization task in this chapter has been defined for a fixed time horizon. If we do not fix the time interval, supposing that the company will operate forever, or for a large period of time, the results can be quite different. In such case we should redefine the optimization criteria because the total revenue at the end of the period will be less adequate. Once this optimization goal is defined, a similar optimization tool can be applied. Further applications should include the multiple product case.

The possibility of viewing the shape of reachable sets provides important information about model behavior. The same simulation program can be used for many similar experiments, looking for reachable sets with variable advertisement and the uncertainty of any other model parameter. Adding more inertia to the model will result in higher state vector dimensionality and more complicated images of the reachable sets. For such cases, the problem is not only the set determination, but also the way to display it. As the computer screen is (still) 2D, good effects can be obtained while displaying the cloud of points in movement, rotating around one of the state-space axes.

The comparison of the DI approach with conventional sensitivity analysis is interesting, but it should be remembered that the problem statement in both approaches is quite different. Most of the sensitivity analysis methods are designed to find the variance propagation and statistical properties of the model output, while DIs and the reachable sets are deterministic. The selection of an appropriate methodology always depends on the purpose of the research.

References

[1] J.P. Aubin, A. Cellina, Differential Inclusions, Grundlehren der Mathematischen Wissenschaften, vol. 264, Springer Verlag, Berlin, Heidelberg, ISBN 978-3-642-69514-8, 1984.

[2] R. Beard, Optimal fertilizer carryover and von Liebigs law of the minimum, in: Conference Paper: Conference 44th, Australian Agricultural and Resource, Economics Society, Sidney, 2000.

[3] D. Bertsimas, A.W. Lo, Optimal control of execution costs, Journal of Financial Markets 1 (1998) 1–50.

[4] A. Brace, D. Gaterek, M. Musiela, The market model of interest rate dynamics, Mathematical Finance 7 (2) (1997) 127–147.

[5] E. Burmeister, A.R. Dobell, Guidance and optimal control of free-market economics: a new interpretation, IEEE Transactions on Systems, Man and Cybernetics SMC-2 (1) (1972) 9–15.

[6] G.C. Chow, Dynamic Economics: Optimization by the Lagrange Method, Oxford University Press, Oxford, ISBN 9780195101928, 1997.

[7] H. Courtney, J. Kirkland, P. Viguerie, Strategy under Uncertainty, Harvard Business Review Nov-Dic 1997, 1997, pp. 2–15.

[8] A.K. Dixit, Optimization in Economic Theory, Oxford University Press, Oxford, 1990.

[9] G. Feichtinger, R.F. Hartl, S.P. Sethi, Dynamic optimal control models in advertising: recent developments, Management Science 40 (2) (1994) 195–226.

[10] M.I. Gomes Salema, A.P. Barbosa-Povoa, A.Q. Novais, An optimization model for the design of a capacitated multi-product reverse logistics network with uncertainty, European Journal of Operational Research 179 (3) (2007) 1063–1077.

[11] I. Karatzas, Optimization problems in the theory of continuous trading, SIAM Journal on Control and Optimization 27 (6) (1989) 1221–1259.

[12] I. Karatzas, J.P. Lehoczky, A.Q. Novais, S.E. Shreve, Optimal portfolio and consumption decisions for a "small investor" on a finite horizon, SIAM Journal on Control and Optimization (ISSN 0363-0129) 25 (6) (1987) 1557–1586, https://doi.org/10.1137/0325086.

[13] W.R. King, Quantitative Analysis for Marketing Management, McGraw-Hill, New York, 1967.

[14] H. Konno, H. Yamazaki, Mean-absolute deviation portfolio optimization model and its applications to Tokyo stock market, Management Science 37 (5) (1991) 519–531.

[15] R. Korn, E. Korn, Option Pricing and Portfolio Optimization Modern Methods of Financial Mathematics, vol. 31, American Mathematical Society, ISBN 0-8218-2123-7, 2001.

[16] E.B. Lee, L. Markus, Foundations of Optimal Control Theory, Wiley, ISBN 978-0898748079, 1967.

[17] M.E. Lien, The virtual consumer; construction of uncertainty in marketing discourse, in: Christina Garsten, Monica Lindh de Montoya (Eds.), Market Matters: Exploring Cultural Processes in the Global Marketplace, Palgrave Publishers, ISBN 1-4039-1757-4, 2004.

[18] G.L. Lilien, P. Kotler, Marketing Decision Making: a Model-Building Approach, Harper & Row, New York, ISBN 0060440767, 1972.

[19] D.B. Montgomery, G.L. Urban, Management Science in Marketing, Prentice Hall, Englewood Cliffs, NJ, 1969.

[20] Felipe León Olivares, Justus Von Liebig: pionero de la enseñanza científica en el campo de la química, in: Memorias del XIII Simposio Estrategias Didácticas en el Aula, Colegio de Ciencias y Humanidades, Universidad Nacional Autónoma de México, 2011, p. 3, (Accessed 18 May 2021).

[21] M.H. Morris, L.F. Pitt, E.D. Honeyutt, Business-to-Business Marketing: A Strategic Approach, Sage, London, ISBN 0-8039-5964-8, 2001.

[22] S. Raczynski, Differential inclusion solver, in: Conference Paper: International Conference on Grand Challenges for Modeling and Simulation, SCS, San Antonio TX, 2002, 2002.

[23] S. Raczynski, Simulation and optimization in marketing: optimal control of consumer goodwill, price and investment, International Journal of Modeling, Simulation, and Scientific Computing 5 (3) (2014).

[24] S. Raczynski, A market model: uncertainty and reachable sets, International Journal for Simulation and Multidisciplinary Design Optimization (ISSN 1779-6288) 6 (A2) (2015).

[25] S. Read, N. Dew, S.D. Sarasvathy, M. Song, R. Wiltbank, Marketing under uncertainty, Journal of Marketing 73 (3) (2009) 1–18.

[26] E. Polak, Computational Methods in Optimization, Academic Press, New York, ISBN 0125593503, 1971.

[27] L.S. Pontryagin, V.G. Boltyanskii, R.V. Gamkrelidze, E.F. Mishchenko, The Mathematical Theory of Optimal Processes, Interscience, ISBN 2-88124-077-1, 1962.

[28] S.J. Shapiro, M. Tadajewski, C.J. Shultz, Interpreting macromarketing: the construction of a major macromarketing research collection, Journal of Macromarketing 29 (2009) 325, https://doi.org/10.1177/0276146709338706.

[29] C. Simos, C.S. Katsikeas, G. Balabanis, M.J. Robson, Emergent marketing strategies and performance: the effects of market uncertainty and strategic feedback systems, British Journal of Management 25 (2) (2014) 145–165.

[30] P. Slovik, Market uncertainty and market instability, IFC Bulletin 34 (2011) 430–435.

[31] M.G. Speranza, A heuristic algorithm for a portfolio optimization model applied to the Milan stock market, Computers & Operations Research 23 (5) (1996) 433–441.

[32] K.E. Stanovich, Discrepancies between normative and descriptive models of decision making and the understanding/acceptance principle, Cognitive Psychology 38 (1999) 349–385, http://www.idealibrary.com.

[33] Z. Yuanguo, A fuzzy optimal control model, Journal of Uncertain Systems 3 (4) (2009) 270–279, http://www.jus.org.uk.

[34] B.P. Zeigler, Theory of Modeling and Simulation, Wiley-Interscience, New York, 1976.

[35] S.K. Zaremba, Sur les équations au paratingent, Bulletin Des Sciences Mathématiques 60 (1936) 139–160.

CHAPTER FOUR

Uncertainty in stock markets

Abstract

Stock market dynamics strongly depends on human factors that introduce uncertain elements. The problem of uncertainty is formulated here in a deterministic way, using differential inclusions as the main modeling tool. This provides reachable sets for the model trajectories, including the possible extreme values of the stock demand and price. The results of several computational experiments are shown, where the uncertainty consists in erroneous or false information about the actual demand. It is pointed out that when treating the uncertain parameters as random, we cannot obtain the real shape of the model's reachable set. Instead, our approach is deterministic and provides the reachable sets for model variables. This may help in financial planning and stock market management.

Keywords

Stock market, Uncertainty, Sensitivity set

4.1. Stock market models and uncertainty

The main topic of this chapter is the uncertainty problem rather than stock market modeling. We use a simple stock market model for short-time market behavior. Models of such kind can provide important information for financial planning and strategic decisions. Recall that the commonly used *system dynamics* (SD) models are mostly continuous and reflect certain global and averaged behavior of possible changes of the system variables. In models of such kind the human factor is strongly simplified. The behavior of a human agent on a stock market is difficult to model and to predict. Consequently, the application of SD modeling methodology to this case, as well as to other systems with human factors, is doubtful. Another possible approach is agent-based simulation (do not confuse this with stock market agents). Anyway, we should remember that not all that happens in the real world can be described by differential equations.

The stock market model itself is not the main topic of this chapter. We use a simple model of a market with only one stock type, known from the literature. For this and similar models, consult Andersen [1], Minsky [16], Levy et al. [14], or Goodwin [9], to mention only some of thousands of publications on stock market behavior. Some more qualitative comments on stock modeling and the use of models can be found in less academic sources, such as Glassman [8].

As for the other sources, let us mention only some important works on stock market dynamics. Note that none of the models used in these publications use differential inclusions (DIs), and the stock market's reachable sets under market uncertainty are shown. Most of the works described in academic sources are based on the time series approach.

Kimoto et al. [12] deal with the buying- and selling-time prediction system for stocks on the Tokyo Stock Exchange. Modular neural networks are used in the model. Some learning algorithms developed by the authors are applied to the Tokyo Stock Exchange price index prediction system. Profit maximization is the main objective of their research.

Lux and Marchesi [15] consider multiagent models of financial markets and discuss the problem of scaling. Recall that economies of scale deal with the reduction in the per-unit cost of production as the volume of production increases. The main point in that article is that financial price characteristics resemble the scaling laws for systems where large numbers of units interact with each other. The authors present a multiagent model of financial markets.

A similar topic is presented in the book of Mandelbrot [18], who brings together his original papers as well as many original chapters specifically written for that book. In the book we can find remarks on the connection between fractals and economical and financial system behavior. Scaling issues are also included.

Bollersev and Mikkelsen [3] discuss financial market volatility problems. They use Monte Carlo simulations to illustrate the reliability of quasimaximum likelihood estimation methods, standard model selection criteria, and residual-based portmanteau diagnostic tests in this context. Their conclusion is that the apparent long-run dependence in US stock market volatility is best described by a mean-reverting fractionally integrated process (Granger [10]).

In the book of Darley and Outkin [5] we can find applications of agent-based modeling to financial markets. It presents a new paradigm, where markets are treated as complex systems whose behavior emerges as a result of interactions between market participants, market institutions, and market rules. Also, simulations, a comparison with behaviors observed in real-world markets (existence of fat tails, spread clustering, etc.), and predictions about possible outcomes of decimalization are included.

Stauffer and Sornette [19] propose a mechanism of "sweeping of an instability," which exhibits reasonable statistics for the distribution of price changes. Another approach to stock market stochastic processes is proposed by Madan and Seneta [17]. They propose a new model for the uncertainty underlying security prices, the Variance Gamma (V.G.) process.

LeBaron et al. [13] present results from an experimental, agent-oriented computer-simulated stock market, which includes artificial intelligence algorithms. Time series from this market are analyzed from the standpoint of well-known empirical features in real markets. The simulated market is able to replicate several of these phenomena, including fundamental and technical predictability, volatility persistence, and leptokurtosis ("fat-tailed risk"). Gavin [7] presents a model where the price of shares in the stock market substitutes for the real interest rate in the determination of aggregate demand.

From newer publications on stock market models and forecasting, let us note the following. Engstrom [6] investigates forecasts of stock market crashes. One of the conclusions is that the forecasting of stock market crashes remains very difficult, even using the best models, at least among those considered by the author. Grinfeld and Cross [11] derive a system of Fokker–Planck equations to model a stock market where hysteretic agents can take long and short positions. They show numerically that the resulting mesoscopic model has rich behavior, being hysteretic at the mesoscale and displaying bubbles and volatility clustering in particular.

Balcilar et al. [2] examine the relationship between the US crude oil and stock market prices using a Markov switching vector error correction model and a monthly dataset from 1859 to 2013. They indicate the high- and low-volatility regimes of the market behavior. Data from the National Bureau of Economic were used in the research.

Chandra [4] discusses web-based software and its use by agents in stock market management. The virtual system described in the paper learns how the trade market works without real money. Several technologies are considered. The system records the history of stock prices and provides graphical information tools to present these data.

The main problem in the modeling of marketing, economic, social, and similar (soft) systems is the lack of exact information. This uncertainty in model data (initial conditions, parameters, external signals, etc.) and even in model structure requires special treatment. The simplest way to get some information about the behavior of a system with uncertainty is to assume some variables to be subject to random changes and to see the resulting model trajectories. A common opinion is that uncertainty can be treated using stochastic models and probabilistic methods. Note, however, that the very essential definition of uncertainty has nothing to do with stochastic processes. An uncertain variable needs not be random. An uncertain variable or parameter has an uncertain value that may belong to some interval or satisfy some restrictions. It may have no probability distribution and other probabilistic properties. The approach to uncertainty treatment proposed here is based on DIs, and is deterministic.

4.2. The model

The model discussed here may appear to be very simple. Note, however, that building stock market models is not the main topic of this chapter. We use this model to show an application of the DI solver. Note also that the simplicity of the model does not imply simplicity of the model's reachable set. This set, even for a small 2D model, may result to be a complicated object.

We consider a simple model of the dynamics of one stock type. First, we consider an ordinary differential equation (ODE) model.

Let p be the current market price and let p_r be the real value of the stock. We will denote by n the current demand of the stock expressed in number of units. Suppose that this demand is the sum of the demand due to the agents being informed about the stock value (n_r) and the demand due to the agents who observe the price increase/decrease rate and do their trading based on predictions of some kind (n_b). The subscript b stands for the "bandwagon effect." This means that the positive or negative price rate attracts increasing or decreasing numbers of agents, respectively.

Moreover, the demand due to erroneous information is denoted by n_e. This is the uncertain component of the demand n.

To find the model equations, observe the following facts. The demand n depends on the difference between the real and the current stock price. This difference should be expressed in relation to the price, so we assume that the demand can be calculated as follows:

$$n_r = A\frac{p_r - p}{p_r}, \tag{4.1}$$

where A is a constant. In some sources the denominator of (4.1) is equal to p and not to p_r, which makes the model more non-linear. However, this makes little difference if the price does not approach zero and changes moderately. This demand, in turn, determines the price growth rate. Thus, we have

$$\frac{dp/dt}{p} = r(t) = Bf(n), \tag{4.2}$$

where $f(n)$ is defined as

$$f(n) = \begin{cases} n \text{ for } n > -I, \\ -I \text{ for } n \leq -I. \end{cases} \tag{4.3}$$

B is a constant and I is the total number of stocks issued. The function f includes a saturation. This means that the surplus of stocks (which results in negative demand) cannot be greater than I.

The component n_b, which determines the "bandwagon effect," depends on the price increase rate. This reaction of the agents is not immediate and is subject to some inertia. We will use here a simple way to represent this:

$$\begin{cases} n_b(s) = G(s)r(s), \\ G(s) = \dfrac{C}{1 + Ts}. \end{cases} \tag{4.4}$$

Here, s is the differentiation operator (Laplace transform variable) and $G(s)$ is the first-order transfer function. Eq. (4.4) implies the following:

$$\frac{dn}{dt} = \frac{(CBf(n) - n_b)}{T}. \tag{4.5}$$

Eqs. (4.2) and (4.5) describe the dynamics of the model. It is a set of two non-linear ODEs that can be easily solved using any continuous simulation tool. In Fig. 4.1 we can see an example of possible changes of the demand during two trading days. The uncertain (erroneous) component n_e was supposed to belong to the interval $[-500, 500]$. Model parameters are as follows:

$$I = 10{,}000, \quad T = 0.004, \quad p_r = 10, \quad A = 1000, \quad B = 0.00007,$$
$$C = 14{,}286, \quad n_e = 0.$$

The value of $T = 0.004$ was chosen to slow down the oscillations and make the example trajectory more illustrative. In real systems, this parameter may be smaller. The initial conditions for the trajectories of Fig. 4.1 and for all other figures were $p(0) = 8$, $n_b(0) = 0$, which means that we start with undervalued stock, which generates a positive demand. To have

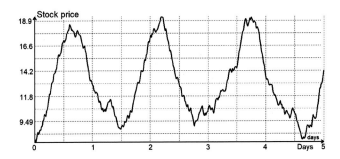

Figure 4.1 Possible changes in demand during five trading days.

a more realistic simulation, a small random disturbance was added to the right-hand side of (4.2). Observe the oscillating pattern of the market trajectory.

4.3. Uncertainty in the stock market

As stated before, the component n_e of the estimated demand represents the erroneous data. A common approach to uncertainty is to treat the uncertain parameters as random. In this case we obtain possible model trajectories and carry out various statistical analyses over sets of hundreds or thousands of trajectories. However, as stated before, the uncertainty should not be confused with randomness. First, to consider a variable as random, you must know that it is really random. In this case, you must know something about its probabilistic properties to be able to generate it. In the case of dynamic systems, not only the probability distribution, but also the spectral density, mutual correlations, etc., are frequently unknown in practical applications.

On the other hand, if a model variable is said to be just uncertain, we only assume some interval (maybe infinite) where it belongs and nothing more. The result of the uncertainty analysis should be the *reachable set* (see Section 1.2.1) for the model variables. Note that such uncertainty treatment is deterministic. This is another reason to treat the uncertain variables in a non-probabilistic way, as such analysis gives us information on possible extreme values (recall the "law of Murphy"). This may also be useful if we expect that the uncertain variables could be intentionally generated to move the system to the extreme values (manipulated and false information).

For example, looking at our model given by Eqs. (4.2) and (4.5), we can see one uncertain variable, n_e (a component of the estimated demand). In vector form our model is described by the following equation:

$$\frac{dx}{dt} = f(x, n_e), \quad n_e \in [-500, 500], \tag{4.6}$$

where x is the state vector ($x = (p, n_b)$) and f is a vector-valued function that includes the two right-hand sides of the equations (for p and n_b). We omit the constant model parameters in the arguments of f. However, n_e appears on the right-hand side of (4.6) because it is a function of time.

Eq. (4.6) can be written as follows:

$$\frac{dx}{dt} \in F(x), \tag{4.7}$$

where F is a set-valued function defined by f, when n_e scans all the values from the interval $[-500, 500]$. What we obtained is a DI instead of a *differential equation*. More details, definitions, and assumptions are given in Chapter 1. What we need as the result of the uncertainty analysis is the reachable set for model variables, and not any particular model trajectory.

In Section 1.5 (Chapter 1), the concept of *functional sensitivity* is defined. This kind of sensitivity supports the perturbations of parameters, fluctuating in time. It is pointed out that the sensitivity is given by the reachable set of the corresponding DI.

In this very natural way, the uncertainty in dynamic system modeling leads to DIs as the corresponding mathematical tool. Note that this tool has been known for about 70 years and that a wide literature body on DI theory and applications is available. In the 1930s, problems such as the existence and properties of solutions to DIs were solved in finite-dimensional spaces. Subsequently, many works on DIs in more abstract, infinite-dimensional spaces appeared. Within a few years after the first publications, DIs resulted to be a fundamental tool in optimal control theory. Recall that optimal trajectories of a dynamic system are those that lay on the boundary of the system's reachable set. This relation to control theory has already been discussed in previous chapters.

4.4. Differential inclusion solver

The determination of the reachable set of a DI is not an easy task. Reachable sets cannot be determined by applying simple perturbations to the system model and looking where the graphs of the model trajectories are located.

The DI solver is described in more detail in Chapter 2. Here we only recall its main features. One could expect that a solution algorithm for a DI may be obtained as some extension of known algorithms for ODEs. However, this is not true. First, the solution to a DI (reachable set) is a set. Namely, it is a set in the time–state space, where the graphs of all possible trajectories of a DI are included. It might appear that the reachable set of a DI can be obtained by simple random shooting, in our case, by generating n_e randomly and then looking for the boundary of the resulting cloud of points reached by the trajectories. Unfortunately, this is not the case, except perhaps for some very simple and trivial models. What we obtain by such primitive random shooting is a cluster of trajectories in a small region that has little to do with the true shape of the reachable set, even with a great number of calculated trajectories. The DI solver scans the boundary of the reachable set instead of its interior. This provides a set of trajectories that can be used to visualize the attainable set. At the final time, the end points of the trajectories form a cloud of points. In the 2D or 3D case and with low dimensionality of the control vector, we can obtain a clear image of the reachable set. In more complicated cases we can only see some projections of that cloud on 2D planes. As stated before, a common error when looking for the shape of a reachable set is to explore its interior. The DI solver used here scans the boundary and not the interior of the set.

4.5. Some results: uncertainty set

4.5.1 Experiment 1: uncertain erroneous information

Here and in the following experiments, the initial demand is equal to zero and the initial price is equal to 8. We examine the reachable set for the model trajectories with the following parameters: the stock surplus (limit of the negative demand) $I = -10{,}000$, $B = 0.00007$, $T = 0.004$, $p_r = 10$, the uncertainty interval for n_e is $[-300, 300]$, $A = 1000$, and $C = 1/B$ (see Section 4.2).

Fig. 4.2 shows the solution (the reachable set) of our DI at the end of one day of trading. The bold contour shows the boundary of the reachable set, that is, the boundary of the set where the model trajectories must belong on the price–demand plane. More precisely, the vertical axis is the value of the n_b component of the demand. This is the demand charged with a first-order inertia, slightly delayed with respect to the real demand.

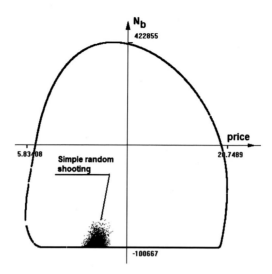

Figure 4.2 Experiment 1. Time section of the reachable set at time $= 1$.

The horizontal axis is the actual price. Note that the demand depends explicitly, in a linear way, on the actual price, so the price-demand plot would not be interesting. The price-n_b plot gives a better insight into the model dynamics.

The contour in Fig. 4.2 was obtained by storing about 1000 model trajectories. To see how useless the primitive random shooting method mentioned before is, the figure also shows the result of such shooting with 10,000 trajectories integrated (a small cluster of pixels inside the reachable set). In this primitive shooting, the random values of n_e were generated on the boundary of the allowed interval $[-500, 500]$. When generating n_e randomly from inside of this interval, the cluster becomes even smaller.

The point is that primitive shooting provides no solution at all. On the other hand, the DI solver is rather slow because of the complexity of the algorithm, which needs the Hamiltonian to be maximized at each integration step. In the presented case, about 5 min of computing time was necessary to get the solution, using a 2Ghz, one processor PC. Fig. 4.3 shows a 3D image of the reachable set.

The lower limit for the demand is clearly seen in the images. The trajectories that determine the reachable set are oscillating around the boundary of the reachable set. These non-linear oscillations suggest that the extreme

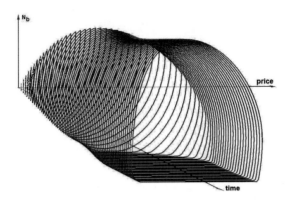

Figure 4.3 3D image of the reachable set for Experiment 1.

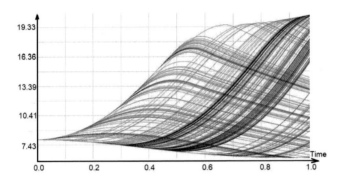

Figure 4.4 Some randomly selected trajectories for Experiment 1. Projection on the n_b-time plane (price).

points of the boundary of the reachable set are reached when the model enters a "resonance" of some kind.

This is hardly possible with a random excitation, but quite possible when the uncertain parameter is changed intentionally to reach the boundary or extreme points. Fig. 4.4 shows some of such randomly selected trajectories. Note that these are not random trajectories. Each of the trajectories of Fig. 4.4 is a 2D projection of a trajectory that scans the boundary of the reachable set.

4.5.2 Experiment 2: slow bandwagon effect

In this experiment, we use the same data as in Experiment 1, except the parameter T, which is now equal to 0.008. This means that the influence

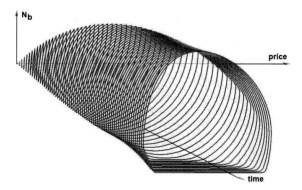

Figure 4.5 The reachable set for Experiment 2.

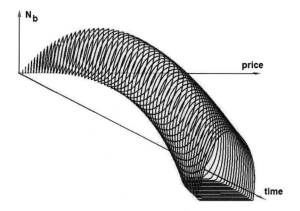

Figure 4.6 The reachable set for Experiment 3.

of the bandwagon effect is two times slower. In Fig. 4.5 we can see the shape of the reachable set in the same coordinates.

4.5.3 Experiment 3: smaller uncertainty range

Now, the data are the same as in Experiment 1, with a smaller uncertainty range of n_e, changed to ± 100. Fig. 4.6 shows the image of the corresponding reachable set. The volume of the reachable set is smaller in this case. The demand saturation is reached at the very end of the simulated interval.

Fig. 4.7 shows the reachable set contours at the final time for Experiments 1, 2, and 3.

Finally, to have an idea about the stability of the model, we run the DI solver with the data of Experiment 1, but with a larger time horizon,

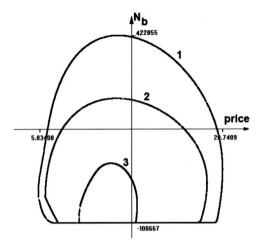

Figure 4.7 Comparison of the reachable sets for Experiments 1, 2, and 3.

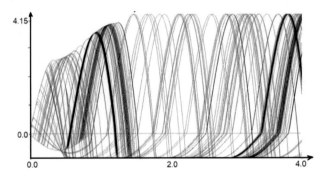

Figure 4.8 Non-linear stable oscillations. Data of Experiment 1 with final time equal to 4.

equal to 4. Fig. 4.8 shows some randomly selected trajectories that scan the boundary of the reachable set. It can be noted that after some initial warm-up period, the trajectories seem to be stable (in control theory this is called orbital stability). The oscillations are not sinusoidal and reveal the non-linearity of the model.

4.6. Conclusion

In this chapter, DIs are applied to the uncertainty problem in dynamic systems. In our approach, the uncertainty problem is deterministic. If we treat the uncertain parameters as random ones and apply simple Monte Carlo-like shooting, we may obtain very poor estimates of the possible extreme values that the model variables can attain due to the uncertainty. The presented DI solver works quite well, though it is still under construction. The stock market dynamics model is a good example of uncertainty and may provide interesting information about possible stock market behavior. The results can be used in financial planning.

The shape of the reachable set may be useful and may tell us more than information obtained by classical sensitivity analysis. For example, looking at Fig. 4.2, the stock market runner can see not only the extreme values; he/she can see, for example, that the combination of stock price equal to 18 and demand equal to 160,000 is possible, while the point (price, demand) = (18, 420) is not.

References

[1] T. Andresen, The dynamics of long-range financial accumulation and crisis, Psychology, and Life Sciences 3 (2) (1999).

[2] M. Balcilar, R. Gupta, S.M. Miller, Regime switching model of US crude oil and stock market prices: 1859 to 2013, Energy Economics (2014), https://doi.org/10.2139/ssrn.2502152, http://ssrn.com/abstract=2502152.

[3] T. Bollerslev, H.O. Mikkelsen, Modeling and pricing long memory in stock market volatility, Journal of Economics 73 (1) (1996) 151–184.

[4] A. Chandra, Developing a model for online stock market trading terminal, Social Science Research Network (2013), https://doi.org/10.2139/ssrn.2264615, http://ssrn.com/abstract=2264615.

[5] V. Darley, A.V. Outkin, Nasdaq Market Simulation: Insights on a Major Market from the Science of Complex Adaptive Systems, World Scientific Publishing Co., Inc., River Edge, NJ, USA, ISBN 9812700013, 2007, 9789812700018.

[6] E. Engstrom, Forecasting Stock Market Crashes Is Hard–Especially Future Ones: Can Option Prices Help?, FEDS Notes, Board of Governors of the Federal Reserve System, 2014.

[7] M. Gavin, The stock market and exchange rate dynamics, Journal of International Money and Finance 8 (2) (1989) 181–200.

[8] K.J. Glassman, Trying to Crack the Code; Computer Models Lag as Stock Pickers, The Washington Post, WP Company LLC d/b/a The Washington Post, October 15, 1998.

[9] R.M. Goodwin, A growth cycle, in: Capitalism and Economic Growth, Cambridge University Press, 1967.

[10] C.W. Granger, An introduction to long-memory time series models and fractional differencing, Journal of Time Series Analysis 1 (1980) 15–30, https://doi.org/10.1111/j.1467-9892.1980.tb00297.x.

[11] M. Grinfeld, R. Cross, A mesoscopic stock market model with hysteretic agents, Discrete and Continuous Dynamical Systems. Series B (ISSN 1531-3492) 18 (2) (2013) 403–415, https://doi.org/10.3934/dcdsb.2013.18.403, http://www.scopus.com/inward/record.url?scp=84874876835&partnerID=8YFLogxK.

[12] T. Kimoto, K. Asakawa, M. Yoda, M. Takeoka, Stock market prediction system with modular neural networks, in: Conference Paper: 1990 IJCNN International Joint Conference on Neural Networks, 1, San Diego, CA, 1990, http://ieeexplore.ieee. org/xpl/opaccnf.jsp.
[13] B. LeBaron, W.B. Arthur, R. Palmer, Time series properties of an artificial stock market, Journal of Economic Dynamics and Control 23 (9–10) (1999) 1487–1516.
[14] M. Levy, H. Levy, S. Sorin, A microscopic model of the stock market: cycles, booms, and crashes, Economics Letters 45 (1) (1994) 103–111.
[15] T. Lux, M. Marchesi, Scaling and criticality in a stochastic multi-agent model of a financial market, Nature 397 (1999) 498–500, https://doi.org/10.1038/17290.
[16] H. Minsky, Can "It" Happen Again? Essays on Instability and Finance, M. E. Sharpe, Inc., Nueva York, 1982.
[17] D.B. Madan, E. Seneta, The variance gamma (V.G.) model for share market returns, The Journal of Business 63 (4) (1990) 511–524, http://www.jstor.org/stable/2353303.
[18] B.B. Mandelbrot, Fractals and Scaling in Finance: Discontinuity, Concentration, Risk, Springer-Verlag, 1997;
H. Minsky, The financial instability hypothesis: an interpretation of Keynes and alternative 'standard theory', in: Recession and Economic Policy, Wheatsheaf, Sussex, 1982.
[19] D. Stauffer, D. Sornette, Self-organized percolation model for stock market fluctuations, Physica. A, Statistical Mechanics and Its Applications 271 (3–4) (1999) 496–506.

CHAPTER FIVE

Flight maneuver reachable sets

Abstract

An application of the differential inclusion solver to a flight dynamics model is discussed. Ordinary differential equations are the main mathematical tool in flight simulation. However, in many situations this is not the best way to solve problems. In robust flight control design, safety, missile and aircraft guidance, analysis of the influence of perturbations, and pursuit–evasion problems, more versatile tools are needed. A *differential inclusion* is a generalization of a differential equation and can be extremely useful. The solution to a differential inclusion is the *reachable set*, and not just a model trajectory or a set of trajectories obtained by randomization of the original problem. In this book, we deal with the reachable sets of differential inclusions related to models. Note that the reachable set is a deterministic object.

Application of the *differential inclusion solver* (Chapter 2) can give a proper view on the regions of the state space where all the possible model trajectories belong. In the present chapter we look for the solutions to the corresponding differential inclusions, i.e., the reachable sets, for flight variables.

Keywords

Differential inclusion, Uncertainty, Flight dynamics, Reachable set

5.1. Differential inclusions and control systems

A survey on differential inclusions (DIs) with more mathematical rigor has already been given in Chapter 1. Here we only recall the most relevant concepts.

An ordinary differential equation (ODE) in a real vector space that can be defined as follows:

$$\frac{dx}{dt} = f(x, t), \tag{5.1}$$

where $x \in R^n$ is the system state vector, f is a vector-valued function, and t is the model time. The *DI* can be treated as a generalization of the above equation. It has the following form:

$$\frac{dx}{dt} \in F(x, t), \tag{5.2}$$

with given initial set for the values of $x(0)$. Here, $F(x, t)$ is a set-valued function $F : R^n \times R \to P(R^n)$ (P means all subsets of R^n).

Reachable Sets of Dynamic Systems
https://doi.org/10.1016/B978-0-44-313384-8.00004-X

As stated in Chapter 1, we make some regularity assumptions about the models considered here. First, we restrict the models to finite-dimensional spaces. Note that one important assumption about the set F of (5.2) is that it changes continuously (sometimes the lower semi-continuity is required) with respect to the Hausdorff distance between sets.

If the set F can be parametrized by a certain variable u, then the DI can be defined in the form of an *equivalent control system* as follows:

$$\begin{cases} x'(t) = f(x, u, t), \\ x(0) \in X_0 \text{ (initial set)}, \\ u \in C(x, t) \ x \in R^n, \ C \subset R^m, \ t \in I, \end{cases} \tag{5.3}$$

where I is a non-empty time interval $[0, T]$ and the prime symbol stands for time differentiation.

A function $z(t)$ such that $z(t) \in F(x, t)$ for all $t \in I$ is called a *selector* of the DI. A *trajectory* $x(t)$ of the DI is a function that satisfies the equation $dx/dt = z(t)$ for almost all $t \in I$. The solution to a DI is not given by any particular trajectory. In this book, by "solution to a DI" we understand the *reachable* or *attainable set* defined as the union of the graphs of all trajectories of the DI. It has been pointed out that with some regularity assumptions, each section of the reachable set by a plane $t = $ const is closed and connected. Each trajectory that reaches the boundary of the reachable set for the given final time T is optimal due to certain optimality criteria. This fact is well known in optimal control theory. Moreover, the whole graph of such trajectories must belong to the boundary of the reachable set. This property can be used to scan the boundary of the reachable set. Such scanning algorithm has been implemented in the *DI solver* described in Chapter 2.

5.2. Application to aircraft maneuvers

An important problem in aircraft control is the resolution of conflict situations. What we need is the shape of the set of possible aircraft positions after a maneuver. In other words, the information that the probability of an accident was equal to 0.0001 is not very relevant to a victim of an accident. A passenger rather wants to know if the accident or collision is possible or not. This information may be important in many other situations as well. For example, the classical missile–plane (pursuit–evasion) game cannot be won by the missile if the reachable sets of the plane and missile positions do not intersect. This information may be used when considering differential games.

Some relevant research on DIs in game theory has been done. Many of the works in the field use the Hamilton–Jacobi–Bellman equations and the methods of control theory closely related to DIs. See for example Tsunumi and Mino [14], who work on the Markov perfect equilibrium problem in differential games. Grigorieva and Ushakov [5] consider the differential game of pursuit–evasion over a fixed time segment. The attainable set is appointed with the help of the stable absorption operator. A more general, variational approach to differential games can be found in Berkovitz [1]. DIs are used by Solan and Wieille [13] to study the equilibrium payoff in quitting games. For general problems of game theory, consult, for example, Petrosjan and Zenkevich [9], Isaacs [7], or Fudenberg and Tirole [4].

The application of DIs in flight control problems is not new. There are many publications on this topic, mainly using DIs as part of an optimal control algorithm. However, that problem is quite different from the application described here. What we are looking for is the shape of the whole reachable set, and not a particular optimal trajectory. Thus, we need the DI solver described in Chapter 2, rather than an optimization algorithm. Though the solver requires multiple solutions of the system state equations and of the conjugated vector trajectory, it is not the same as solving a series of dynamic optimization tasks. Examples of applications of DIs in optimal control can be found mainly in the *Journal of Guidance, Control, and Dynamics*. For example, Seywald [12] describes an optimization algorithm based on DIs. A similar problem for desensitized optimal control is described in [11].

An extensive and detailed report on an application of DIs in flight control is given by Dutton [2]. In that paper, the problem of the existence of a return-to-launch-site trajectory for a space vehicle is considered. This is an important topic in abort mission scenarios, and is related to the determination of reachable sets. Other applications of DIs to optimal control can be found in Mordukhovich [8], Raivio [10], and Fahroo and Ross [3].

5.3. Flight dynamics

The model we use is given by the following set of five differential equations, which describe simplified flight dynamics (see Figs. 5.1 and 5.2).

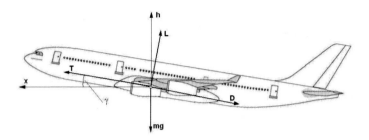

Figure 5.1 Aircraft side view.

It is supposed that the aircraft is in on-route flight. The influence of the wind is neglected. The angle of attack and the flight path angles are assumed to be small. The thrust and drag are supposed to be aligned. The initial value of the inertial heading is set equal to 90 degrees. The fuel consumption is neglected during the maneuver and the air density is supposed to be constant. Consult Hull [6]. We have

$$
\begin{cases}
v'(t) = \dfrac{(T - D)}{m} - g\gamma, \\
h'(t) = v\gamma, \\
\Psi' = \dfrac{L\sin(\varphi)}{mv}, \\
x' = v\sin(\Psi), \\
y' = v\cos(\Psi), \\
L\cos(\varphi) = mg,
\end{cases}
\tag{5.4}
$$

where T is the thrust, D is the drag, L is the lift, φ is the aerodynamic bank angle, γ is the flight path angle, Ψ is the inertial heading, and (x, y) indicates the position (with the positive x-axis indicating the forward direction). Variables T, γ, and φ are not exactly defined; only restrictions for these variables are given. Thus, when T, γ, and φ scan their limits, the right-hand side of (5.4) scans a set in the 5D state space. In this way we obtain a DI, whose solution (reachable set) shows the possible range of the flight state variables. The drag D is calculated as follows.

First, the coefficient C_L is calculated from the equation

$$
L = \frac{1}{2}\rho v^2 S C_L,
\tag{5.5}
$$

where ρ is the air density, S is the gross wing area, and L is the lift.

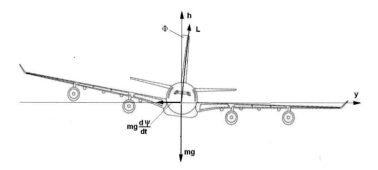

Figure 5.2 Aircraft front view.

The drag coefficient C_D is calculated from the formula $C_D = C_0 + C_{DT}C_L^2$, where C_0 and C_{DT} are known from the aircraft data. Finally, the drag is calculated as follows:

$$D = \frac{1}{2}\rho v^2 S C_D. \tag{5.6}$$

In the experiments shown here, the following parameter values have been assumed:

$C_0 = 0.018$,
$C_{DT} = 0.0342$,
$m = 200,000$ kg,
$S = 363$ m^2,
$T = 1,600,000$ N,
$\rho = 0.5$ kg/m^2,
$\gamma \in [-18°, +18°]$,
$\varphi \in [-21°, +21°]$,
$v(0) = 200$ m/s,
$h(0) = 0$ (relative height),
$\Psi(0) = 90°$,
$x(0) = 0$,
$y(0) = 0$.

In the following figures, the state variables are as follows:

$$X_1 = v, \ X_2 = h, \ X_3 = \Psi, \ X_4 = x, \ X_5 = y.$$

Fig. 5.3 shows the reachable set for the flight trajectories and the y- and h-coordinates. These are the trajectory end points reached after 15 s of

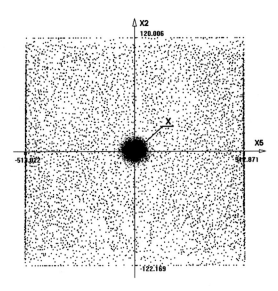

Figure 5.3 The reachable set for time = 15, *y*- and *h*-coordinates.

flight. Each dot represents a position that belongs to the boundary of the reachable set. Some points seem to be inside the reachable set, but this is not true. Note that what we can see on a 2D image is only a projection from a 5D set of points. Observe the small cluster of points (small dots) in the center of the figure. This set is the result of simple random shooting, where the control variables (γ and φ) are generated as random ones, within the same limits. The cluster contains 10,000 trajectory end points, while the reachable set obtained with the DI solver consists of only 5000 trajectories. It can be seen how inefficient the simple (primitive) random search is.

In Fig. 5.4 we see the 3D image of the same reachable set, shown as a cloud of points, from two different view angles. In the right image, bold lines have been manually added to make the shape of the cloud more visible.

If we increase the limits for the control variables and the final flight time, the reachable set becomes very deformed due to the non-linearity of the model. In Fig. 5.5 we see an image similar to that in Fig. 5.3, with the final flight time equal to 30 s and $\gamma \in [-36°, +36°]$. In the static image, the 3D shape may be not so clear. The DI solver displays these images rotating around the selected axis, so the shape is clearly seen.

The system trajectories are stored in a file, together with the respective controls. Thus, selecting any point from the reachable set image we can see not only the model trajectory, but also the corresponding strategies

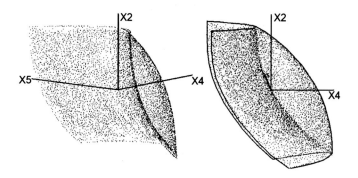

Figure 5.4 3D images of the reachable set.

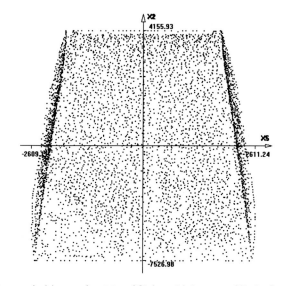

Figure 5.5 The reachable set after 30 s of flight, with increased limits for the flight path angle.

(controls). Fig. 4.6 shows a 3D view of the reachable set with the same model parameters as for Fig. 5.5. See Fig. 5.6.

5.4. Conclusion

The DI solver works quite well, though further research is necessary. The main problems in developing DI solvers are the computational complexity and the visualization of the results in the multidimensional case (see the conclusions to Chapter 3).

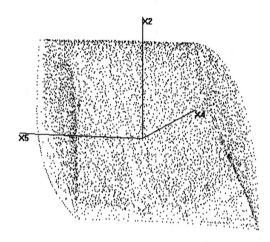

Figure 5.6 3D view of the reachable set with the same model parameters as for Fig. 5.5.

As for the flight simulations, in many situations knowledge about the reachable set may be important in safety, parameter uncertainty analysis, and robust control design. The new implementations of the solver with multiprocessing may be fast enough to provide pilots with additional information, running in real time.

References

[1] L.D. Berkovitz, A variational approach to differential games, in: L.S. Shapley, A.W. Tucjer (Eds.), Advances in Game Theory, Princeton University Press, Princeton, NJ, 1964.
[2] K.E. Dutton, Optimal Control theory Determination of Feasible Return-to-Launch-Site Aborts for the HL-20 Personnel Launch System Vehicle, Report NASA Technical Paper 3449, Langley Research Center, 1994.
[3] F. Fahroo, I.M. Ross, Second look at approximating differential inclusions, Journal of Guidance, Control, and Dynamics 24 (1) (2001) 131–134.
[4] D. Fudenberg, J. Tirole, Game Theory, MIT Press, 1991.
[5] S.V. Grigorieva, V.N. Ushakov, Use finite family of multivalued maps for constructing stable absorption operator, Topological Methods in Nonlinear Analysis 15 (1) (2000).
[6] D. Hull, Fundamentals of Airplane Flight Mechanics, Springer, Berlin, ISBN 3-540-46571-5, 2007.
[7] R. Issacs, Differential Games, Dover Publications, Inc., New York, 1999.
[8] B.S. Mordukhovich, Optimal Control of Nonconvex Differential Inclusions. Differential Equations, Chaos and Variational Problems, Springer, 2007, pp. 285–303.
[9] L.A. Petrosjan, N.A. Zenkevich, Game Theory, World Scientific Publishing Co., Inc., London, 1996.
[10] T. Raivio, Computational Methods for Dynamic Optimization and Pursuit-Evasion Games, Research Reports, A90, Helsinki University of Technology, Systems Analysis Laboratory, 2000.

[11] H. Seywald, Desensitized optimal trajectories with control constraints, in: Conference Paper: Proceedings of the AAS/AIAA Space Flight Mechanics Meeting, Paper no. AAS 03-147, Ponce, Puerto Rico, 2003.

[12] H. Seywald, Trajectory optimization based on differential inclusions, Journal of Guidance, Control, and Dynamics 17 (3) (1994) 480–487.

[13] E. Solan, N. Vieille, Quitting games, Mathematics of Operations Research 26 (2001) 265–285.

[14] S. Tsunumi, K. Mino, Nonlinear strategies in dynamic duopolistic competition with sticky prices, Journal of Economic Theory 52 (1990) 136–161.

CHAPTER SIX

Vessel dynamics and reachable sets

Abstract

In robust flight control, surface or underwater ship control design, safety, missile and aircraft guidance, analysis of the influence of disturbances, and differential games, differential inclusions may provide new insight into the model dynamics. A *differential inclusion* is a generalization of a differential equation.

Application of the *differential inclusion solver* described in this book can give us a proper view of the regions in the state space where all the possible model trajectories belong. Using a known model of vessel dynamics, the attainable sets for the ship's position and other parameters are obtained using the differential inclusion solver.

Keywords

Differential inclusion, Uncertainty, Ship maneuvers, Reachable set

6.1. Differential inclusions and control systems

A survey on differential inclusions (DIs) with more mathematical rigor has already been given in Chapter 1. For convenience, let us recall here the most relevant concepts.

An ordinary differential equation (ODE) in a real vector space can be defined as follows:

$$\frac{dx}{dt} = f(x, t), \qquad (6.1)$$

where $x \in R^n$ is the system state vector, f is a vector-valued function, and t is the model time. The *DI* can be treated as a generalization of the above equation. It has the following form:

$$\frac{dx}{dt} \in F(x, t), \qquad (6.2)$$

with given initial set for the values of $x(0)$. Here, $F(x, t)$ is a set-valued function $F : R^n \times R \to P(R^n)$ (subsets of R^n).

As stated in Chapter 1, we make some regularity assumptions about the models considered here. First, we restrict the models to finite-dimensional spaces. Note that one important assumption about the set F of (6.2) is that it changes continuously (sometimes the lower semi-continuity is required) with respect to the Hausdorff distance between sets.

Suppose that we can parametrize the set F by a parameter u. Then we can define the DI in the form of an *equivalent control system* as follows:

$$\begin{cases} x'(t) = f(x, u, t), \\ x(0) \in X_0 \text{ (initial set)}, \\ u \in C(x, t) \ x \in R^n, \ C \subset R^m, \ t \in I, \end{cases} \quad (6.3)$$

where $I = [0, T]$ is a non-empty time interval and the prime symbol stands for time differentiation.

A function $z(t)$ such that $z(t) \in F(x, t)$ for all $t \in I$ is called a *selector* of the DI. A *trajectory* $x(t)$ of the DI is a function such that $dx/dt = z(t)$ for almost all $t \in I$. In some sources, a trajectory of a DI is called a solution to the DI. However, in this book, by "solution to a DI" we understand the *reachable* or *attainable set*, defined as the union of the graphs of all trajectories of the DI. It has been pointed out that with some regularity assumptions, each section of the reachable set by a plane $t = \text{const}$ is closed and connected. Each trajectory that reaches the boundary of the reachable set for the given final time T is optimal, according to certain optimality criteria. It has also been pointed out that the whole graph of such trajectory must belong to the boundary of the reachable set. We use this fact to scan the boundary of the reachable set, as implemented in the DI solver described in Chapter 2.

6.2. The model of movement

Ship dynamics is typically represented by a 6D rigid body model. More simplified models that provide the ship transfer function are based on the Nomoto [13] approach. This is a very simplified, but easy to use and practical model used in ship control system design. The fourth-order transfer function is sometimes simplified to second- and first-order models. The popularity of the first-order Nomoto model is due to its simplicity and relative accuracy. It is frequently used in ship steering autopilot design.

To simulate ship movement with higher accuracy, computational fluid dynamics methods are used. Such simulations require tools that resolve the partial differential equations of fluid flow and its interaction with the vessel. However, rigid body dynamics, expressed by ODEs, may provide a quite satisfactory approximation. Let us mention a few sources related to the topic, selected from a wide body of available literature.

Analysis of ship dynamics also involves the problem of stability. Gourlay and Lilienthal [3] propose a method for evaluating the overall dynamic stability of a vessel in a seaway with certain loading and wave conditions. The method can serve as an important tool, providing safe headings in extreme conditions and preserving vessel stability. Ibrahim and Grace [4] also deal with ship behavior in severe environmental conditions. Coupled non-linear equations of movement are presented. The stochastic stability and the probability approach are used to assess the probability of capsizing.

A good classical approach to the parameter identification problem is given in the paper of Astrom and Kallstrom [1]. The maximum likelihood method is used to determine the parameters of a linear model, using discrete-time measurements. Nguyen et al. [12] use a linear quadratic Gaussian control algorithm with an autoregressive ship model. The simulations are carried out to assess the robustness of a ship autopilot system.

Kreuzer and Pick [6,7] deal with ship dynamics and stability. It is pointed out that capsizing of a ship in regular waves is the result of a sequence of bifurcations in the ship's motion, and the determination of bifurcations is possible using path-following techniques of non-linear dynamics. Miyata et al. [10] present an application of computational fluid dynamics methods to the problem of ship interaction with non-linear waves. The code has been developed in order to deal with wave problems based on a rectangular grid system, as well as on the framework of a boundary-fitted grid system.

Oskin et al. [14] present an application of neural networks to the problem of ship parameter identification. They show that even a simple recurrent neural network model can be successfully trained for both linear and non-linear behavior of a ship. Experiments carried out with data taken from a simulator of the ship also confirmed the effectiveness and usefulness of this approach.

A model of a container ship and multiple floating vessels can be found in the paper of Kang and Kim [5]. It is pointed out that prediction of the relative motions of the multiple floating bodies with a realistic mooring system is essential for harbor maneuvers. Both collinear and non-collinear wind–wave–current environments are applied to the system.

Spyrou and Thompson [15] provide a review of the field of non-linear ship dynamics and demonstrate the importance of the subject in the context of naval architecture. In the book of Milward [9] we find a great review of ship stability problems, including stability criteria, wind heel criteria, damaged stability, and dry docking and grounding.

The paper of Tzeng and Chen [16] is focused on first- and second-order Nomoto models. Problems of observability and controllability are discussed. Model reduction for a fourth-order transfer function ship model describing the sway–yaw–roll dynamics is conducted to obtain a second-order Nomoto model.

Understanding of ship dynamics and the possibility to quickly display and use the results is important in emergency maneuvers. It is a good question how Titanic's first officer William Murdoch should have reacted when he knew about the iceberg. Perhaps he could have avoided the collision by turning immediately without reversing the engines. Anyway, immediate information about the reachable positions of the vessel could help. Perhaps there were some parts of the reachable set for the ship's position that did not intersect with the iceberg area.

The ship dynamics model itself is not the main topic of this chapter. Anyway, the model used here is not trivial. It contains important non-linearities and coupled non-linear differential equations. Here, we use a simplified 6D model of ship dynamics to simulate ship motion. The water current is neglected. It is also supposed that the ship's helm does not work when the thrust is equal to zero. The model is given by the following equations:

$$\begin{cases} x' = u\cos(\Theta) - v\sin(\Theta), \\ y' = u\sin(\Theta) + v\cos(\Theta), \\ \Theta' = r, \\ u' = \dfrac{1}{m}\left(T_u + X_u + vr + K_1\cos(\Psi - (\Theta + \pi))W^2\right), \\ v' = \dfrac{1}{m}\left(T_v + Y_v - ur + K_2\sin(\Psi - (\Theta + \pi))W^2\right), \\ r' = \dfrac{1}{I}\left(N_r r - bT_v\right), \end{cases} \qquad (6.4)$$

where m is the ship's mass, I is the moment of inertia, T is the thrust, X_u, Y_v, and N_r are the hydrodynamic coefficients (which are always negative), W is the wind magnitude, Ψ is the wind direction, K_1 and K_2 are (negative) constants, (x, y) is the ship's (absolute) position, u and v are the velocities in ship coordinates (forward and sideways), X_s and Y_s are surge and sway, respectively, and r is the rotation speed (yaw Θ). See the corresponding vectors in Fig. 6.1. In this simplified model, the sea current is neglected. For other, more complicated models, consult Bertin et al. [2] and Nagchaud-

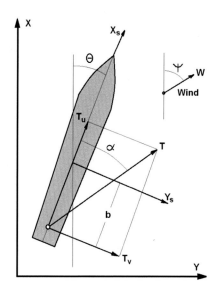

Figure 6.1 Ship coordinates and forces.

huri et al. [11]. Here we use a model similar to that of Nagchaudhuri. See also Matusiak [8].

Let us see some reachable sets of the model (6.4). We treat (6.4) as a control system, where the total thrust T, u, and v are subject to restrictions.

When these variables scan all the permissible values, the right-hand sides of (6.4) scan a set F of (6.2).

This defines a DI. The *DI solver* is used to calculate the corresponding reachable sets. The solver algorithm is described in Chapter 2. Here, only recall that the solver generates a set of trajectories of the DI, scanning the boundary, and not the interior of the reachable set. This is not "random or primitive shooting." We do not just apply simple perturbations within the given restriction set. The trajectories generated by the solver obey the Jacobi–Hamilton equations. Some results from the optimal control theory are used. These trajectories permit us to see the shape of the reachable set.

One of the additional results of the solver is the possibility of retrieving the control variables for any trajectory that reaches the boundary of the reachable set at the final simulation time. The user can select a point in the reachable set final-time section, and the solver shows the corresponding control.

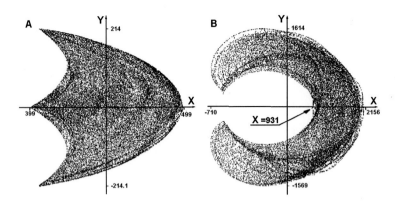

Figure 6.2 The reachable sets for a ship maneuver for time = 40 (part A) and time = 120 (part B).

The shape of the reachable set in the maneuvering problem provides valuable information. For example, we can see if the vessel can avoid collision with an obstacle at all. Of course, similar information might be useful in harbor maneuvering. Solving the DIs for two objects, we can see if it is possible to intercept an enemy object in a pursuit–evasion game or if there is a risk of collision.

In the following simulation experiments, the controls are the thrust T and the angle α (see Fig. 6.1). The right-hand sides in (6.4) define a 6D set, being the mapping from a 2D rectangle on the plane (T, α) to the 6D region with coordinates (x, y, Θ, u, v, r).

6.3. The reachable sets

In our experiments, we use the following parameters: ship mass = 800,000 kg, total thrust $T \in [0, 100,000]$ N, $\alpha \in [-0.471, 0.471]$ rad, $X_u = -200$, $Y_v = -1000$ Ns/m, $b = 15$ m, $I = 170,000,000$ kg m^2/s, and $N_r = -15,000,000$ ms/rad (supposing the angles in (6.4) are in radians). The initial speed X_s is set to 10 m/s, $Y_s = 0$, and other initial conditions are also set to zero. In Fig. 6.2, X and Y are in meters. It is also assumed that the thrust cannot be negative.

Fig. 6.2A shows the intersection of the reachable set with the plane time = 40 s. In Fig. 6.2B we can see a similar image for time = 120 s. Note the big difference in the scale of the two images. Obviously, the reachable set for the larger time interval is greater. Also note that positions near the

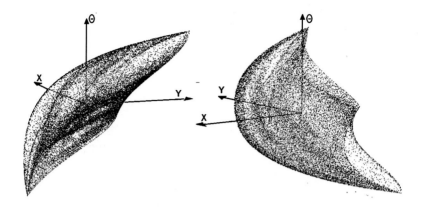

Figure 6.3 3D image of the reachable set for time $= 4$, viewed from different angles.

X-axis are not attainable for X less than 931. This is because the initial speed in the X-direction is equal to 10 m/s, the hydrodynamic coefficient X_u is relatively small, and the thrust cannot be negative. In this model it is also assumed that the ship's rudder does not work when the thrust is equal to zero.

In the following figures, projections of clouds of points into the XY-plane are shown. Note that all these points are on the boundary of the reachable set; they are not internal points. What we see is the surface of a 6D "balloon."

Fig. 6.3 depicts a 3D image of the reachable set for time $= 40$ s. The co-ordinates are X, Y, and the yaw Θ. The interpretation of such images is not easy, and requires certain 3D imagination. Perhaps a specialist in ship maneuvering may get a clear and useful idea about possible ship movements.

In Fig. 6.4 we can see the reachable set for time equal to 40 s, in the presence of wind. The amount of wind is $W = 22$ m/s, with angle $\Psi = 30$ degrees, and coefficients K_1 and K_2 are equal to -20 and -80, respectively. It can be seen that the available ship positions are shifted towards the positive values of Y and the reachable region in the X-direction increases (the ship is being pushed forwards).

Finally, let us see the reachable set for the same model with initial velocity equal to zero. The projection on the XY-plane is shown in Fig. 6.5. Observe the X and Y scales, which are different from those in the previous images. Logically, in this case the point $(x, y) = (0, 0)$ is attainable, unlike in the previous cases.

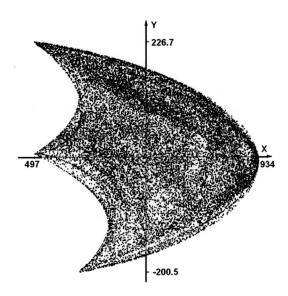

Figure 6.4 Reachable set for time = 40, with influence of the wind.

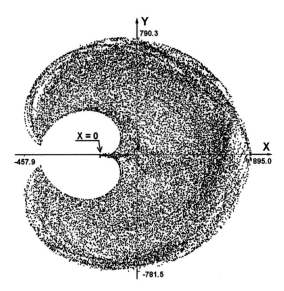

Figure 6.5 Reachable set for time = 120, with initial velocity zero.

6.4. Conclusion

Here, we show only some selected images. The DI solver produces a lot of reachable set projections and 3D images that can be used by specialized crew members (the ship's helmsmen). Knowledge about the reachable set can be important for harbor maneuvering and for safety issues. As mentioned before, the information about the obstacle (e.g., an iceberg) and the reachable set can indicate if it is possible or not to avoid a collision, and how to control the ship.

The reachable set images for models of order higher than two may be somewhat complicated. Further improvements to the software should be made to display these multidimensional point clouds.

References

[1] K.J. Astrom, C.G. Kallstrom, Identification of ship steering dynamics, Automatica 12 (1976) 9–22.

[2] D. Bertin, S. Bittani, S. Meroni, S.M. Savaresi, Dynamic positioning of a single truster vessel by feedback linearization, in: Conference Paper: Proceedings of the 5th IFAC Conference on Manoeuvring and Control of Marine Craft, Aalborg, Denmark, 2000.

[3] T. Gourlay, T. Lilienthal, Dynamic stability of ships in waves, in: Conference Paper: Pacific 2002 International Maritime Conference, Sydney, 2002.

[4] R.A. Ibrahim, I.M. Grace, Modeling of ship roll dynamics and its coupling with heave and pitch, Mathematical Problems in Engineering 2010 (2010), https://doi.org/10.1155/2010/934714.

[5] H.Y. Kang, M.H. Kim, Hydrodynamic interactions and coupled dynamics between a container ship and multiple mobile harbors, Ocean Systems Engineering 2 (3) (2012) 217–228, https://doi.org/10.12989/ose.2012.2.3.217.

[6] E. Kreuzer, M.A. Pick, Dynamics of ship–motion, Proceedings in Applied Mathematics and Mechanics 3 (1) (2003) 84–87, https://doi.org/10.1002/pamm.200310322.

[7] E. Kreuzer, M.A. Pick, Fishing vessel dynamics and stability, in: Conference Paper: Ship Technology and Research (STAR) Symposium, THE International Community for Maritime and Ocean Professionals, 1986.

[8] J. Matusiak, Two stage approach to determination of non-linear motions of ship in waves, in: Conference Paper: 4th Osaka Colloquium on Seakeeping Performance of Ships, Osaka, 2000.

[9] J. Milward, Ship Stability and Dynamics, AMC Search, Launceston, Tasmania, ISBN 0958185077, 2004.

[10] H. Miyata, H. Orihara, Y. Sato, Nonlinear ship waves and computational fluid dynamics, Proceedings of the Japan Academy. Series B Physical and Biological Sciences 90 (8) (2014) 278–300, https://doi.org/10.2183/pjab.90.278.

[11] A. Naghauhuri, P. Hitchener, D. Lous, et al., Integration of the state of art simulation software tools for guidance and control of an under-actuated surface autonomous vessel, in: Conference Paper: Proceedings of the 2004 American Society for Engineering Education Annual Conference, 2004.

[12] H.D. Nguyen, D.M. Le, K. Ohtsu, Ship's optimal autopilot with a multivariate autoregressive exogenous model, in: Conference Paper: 11th IFAC Workshop on Control Applications of Optimization, International Federation of Automatic Control, St. Petersburg, Russia, 2000.

[13] K. Nomoto, K. Taguchi, K. Honda, On the steering quality of ships, International Shipbuilding Progress 4 (1957) 354–370.

[14] D.A. Oskin, A. Dyda, V. Markin, Neural network identification of marine ship dynamics, in: Conference Paper: 9th IFAC Conference on Control Applications in Marine Systems (2013), Osaka, Japan, 2013.

[15] K.J. Spyrou, J.M.T. Thompsom, The nonlinear dynamics of ship motions: a field overview and some recent developments, Philosophical Transactions - Royal Society. Mathematical, Physical and Engineering Sciences 358 (1171) (2000) 1735–1760.
[16] C.Y. Tzeng, J.F. Chen, Fundamental properties of linear ship steering dynamic models, Journal of Marine Science and Technology 7 (2) (1999) 79–88.

CHAPTER SEVEN

Mechanical systems: earthquakes and car suspensions

Abstract

In this chapter, uncertain variable disturbances in models of mechanical systems are discussed. The attainable sets for the state variables are calculated using the differential inclusion solver. A car suspension model and a building's response to earthquake movement are used as examples. In contrast to the commonly used statistical methods, the determination of the reachable sets for the model trajectories using differential inclusions is a deterministic approach.

Keywords

Differential inclusion, Uncertainty, Car suspension, Mechanical system, Reachable set

7.1. Differential inclusions

For the reader's convenience, we repeat here some basic concepts. Consult Chapter 1 for a wider overview of differential inclusions (DIs). Recall that a *DI* is a generalization of a differential equation. It has the following form:

$$\frac{dx}{dt} \in F(x, t), \tag{7.1}$$

with given initial set X_0, $x(0) \in X_0$. Here, $F(x, t)$ is a set-valued function (see Chapter 1 for more details). If the set F includes only one point, the DI "degenerates" to an ordinary differential equation (ODE).

If the set F can be parametrized by a certain variable u, then the DI can be defined in the form of an equivalent control system as follows:

$$x'(t) = f(x, u, t), \quad u \in C(x, t), \tag{7.2}$$

where $f = (f_1, f_2, ..., f_n)$, x is the state vector $x = (x_1, x_2, ..., x_n)$, and u is the control variable $u = (u_1, u_2, ..., u_m)$. The prime mark stands for time differentiation. C represents the limitations for the control u. We suppose that the time runs over a non-empty interval $I = [0, T]$, and the initial condition is given as $x(0) = X_0$ (X_0 reduced to one point).

Reachable Sets of Dynamic Systems
https://doi.org/10.1016/B978-0-44-313384-8.00006-3

119

The relation between the set F of (7.1) and the function f of (7.2) is as follows:

$$F(x, t) = \{z : z = f(x, u, t) \mid u \in C(x, t)\}. \qquad (7.3)$$

A function $x(t)$ that satisfies (7.1) a.e. over a given time interval is called a trajectory of the DI. The solution to a DI is not given by any particular trajectory. In this book, by "solution to a DI" we understand the *reachable* or *attainable set*, defined as the union of the graphs of all trajectories of the DI.

7.2. The differential inclusion solver

The determination of the reachable set of a DI is not an easy task. Reachable sets cannot be determined by applying simple perturbations to the system model and looking where the graphs of the model trajectories are located.

The DI solver has been described in more detail in Chapter 2. Here, we recall its main features. One could expect that a solution algorithm for a DI may be obtained as some extension of known algorithms for ODEs. However, this is not true. First, the solution to a DI (reachable set) is a set. Namely, it is a set in the time-state space, where the graphs of all possible trajectories of a DI are included.

One could expect that the reachable set of a DI can be obtained by a simple random shooting, randomly generating trajectories, and then looking for the boundary of the resulting cloud of points reached by the trajectories. Unfortunately, this is not the case, except perhaps in some very simple and trivial cases. What we obtain by such primitive random shooting is a cluster of trajectories in a small region that has little to do with the true shape of the reachable set, even if a great number of trajectories are calculated.

The DI solver scans the boundary of the reachable set instead of its interior. This provides a set of trajectories that can be used to visualize the attainable set. At the final time, the end points of the trajectories form a cloud of points. In the 2D or 3D case and with low dimensionality of the control vector, we can obtain a clear image of the reachable set. In more complicated cases we can only see some projections of that point cloud on a 2D plane. As stated before, a common error when looking for the shape of the reachable set is to explore its interior. The DI solver used here scans the boundary and not the interior of the set.

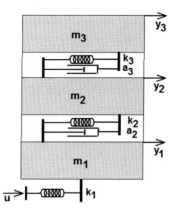

Figure 7.1 The mechanical scheme of a structure.

7.3. An earthquake

In this section we use the DI solver to obtain the reachable sets of the movement of a structure. This provides a wide look at the model behavior, and permits to see the worst possible cases of earthquake movements.

We will not review the large amount of literature available in the field. Let us mention only the article of Catering et al. [4] presenting the methods of building stiffness evaluation. The article of Dimig et al. [7] describes a study of earthquakes based on computer simulation. A good survey of other works can be found in McCrum and Williams [9].

Let us consider a small three-floor building. We will simulate 1D horizontal excitation and the building response in horizontal movement. The mechanical scheme is shown in Fig. 7.1. The earthquake movement is denoted as u. It produces a force that moves the mass m_1 through a spring with coefficient k_1 (N/m). This spring represents the ground elasticity. The equations of movement are as follows (the prime mark means time differentiation):

$$\begin{cases} \dfrac{d^2 y_1}{dt^2} = \dfrac{1}{m_1} \left(k_1(u - y_1) - k_2(y_1 - y_2) - a_2(y_1' - y_2') \right), \\[3mm] \dfrac{d^2 y_2}{dt^2} = \dfrac{1}{m_2} \left(k_2(y_1 - y_2) + a_2(y_1' - y_2') - k_3(y_2 - y_3) - a_1(y_2' - y_3') \right), \\[3mm] \dfrac{d^2 y_3}{dt^2} = \dfrac{1}{m_3} \left(k_1(u - y_1) - k_2(y_1 - y_2) - a_2(y_1' - y_2') \right). \end{cases}$$

$$(7.4)$$

In order to get a system of equations of the first order, we introduce the following state variables: $x_1 = y_1$, $x_2 = y_2$, $x_3 = y_3$, $x_4 = y_1'$, $x_5 = y_2'$, and $x_6 = y_3'$. The state equations are as follows:

$$\begin{cases} x_1' = x_4, \\ x_2' = x_5, \\ x_3' = x_6, \\ x_4' = \dfrac{1}{m_1}\left(k_1(u - x_1) - k_2(x_1 - x_2) - a_2(x_4 - x_5)\right), \\ x_5' = \dfrac{1}{m_2}\left(k_2(x_1 - x_2) + a_2(x_4 - x_5) - k_3(x_2 - x_3) - a_3(x_5 - x_6)\right), \\ x_6' = \dfrac{1}{m_3}\left(k_3(x_2 - x_3) + a_3(x_5 - x_6)\right). \end{cases}$$

$$(7.5)$$

We use the following model parameters: $m_1 = 200{,}000$ kg, $m_2 = 200{,}000$ kg, $m_3 = 100{,}000$ kg, $k_1 = 4 \times 10^4$ kN/m, $k_2 = 2 \times 10^4$ kN/m, and $k_3 = 1.5 \times 10^4$ kN/m.

The DI solver may fail if there are big differences between model parameters. Thus, we change the scale as follows: $m_1 = 2$, $m_2 = 2$, $m_3 = 1$, $k_1 = 400$, $k_2 = 200$, and $k_3 = 150$.

In fact, this is only a change of the time scale. The natural frequencies of the model remain the same as before. The damper coefficients have been defined as follows: $a_2 = 6$, $a_3 = 5.2$. With these parameters of the shock absorbers the value of the corresponding standard oscillator damping coefficient is about 0.15. Anyway, we are looking rather for qualitative and not quantitative results, so the parameters are defined as typical ones, and not taken from a real building.

In Fig. 7.2 we can see the results of a simple simulation of the model. The excitation u was a step function, $u \equiv 0.5$ m $\forall t > 0$. Thus, the whole structure moves 0.5 m to the right, with some oscillations. Note that the amplitude of the first (the greatest) oscillation of the upper mass is approximately equal to 1.13. The movements of the three floors are oscillatory, being a superposition of several different frequencies.

Now, we treat the model equations in (7.5) as a DI. The right-hand side of the equation is treated as a set-valued function, where the uncertain variable u (earthquake movements) scans all possible values within given limits. Note that now u is not a constant parameter, but it changes in time.

Fig. 7.3 shows the shape of the reachable set, the time section for time $= 3.5$ s (coordinates y_1 and y_3). Remember that all displayed points are the

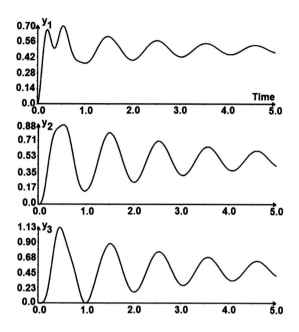

Figure 7.2 Simple simulation of the model (7.5).

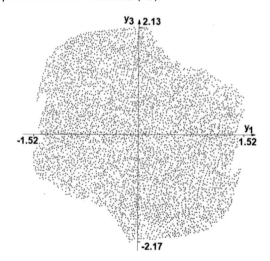

Figure 7.3 The reachable set for model (7.5) at time = 3.5.

end points of the trajectories that belong to the boundary of the reachable set. Some points appear to lie inside the set because this is a projection of a 6D cloud of points onto a 2D plane. The excitation u for this simulation

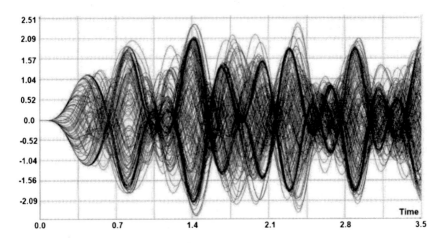

Figure 7.4 Some boundary scanning trajectories.

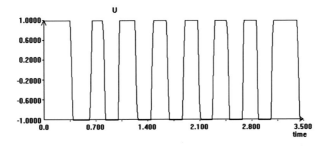

Figure 7.5 The form of the earthquake movement that results in the biggest amplitude of the structure movement.

is a function of time, oscillating between ±0.5 m. The uppermost point of the point cloud is the end point of the trajectory that reaches the biggest value of y_3, achieved at time = 3.5. Observe that the value is equal to 2.13, i.e., the movement is considerably greater than the amplitude that results from a simple simulation (Fig. 7.2) with constant u. Fig. 7.4 shows some (1 out of each 20) boundary scanning trajectories as a function of time. Note that all trajectories oscillate. This means that each trajectory slides over the boundary of the reachable set in a spiral-like movement.

Let us check the optimal control for the trajectory with the greatest value of y_3 at time = 3.5. This control represents the worst-case movement of the earthquake that may destroy the structure. The DI solver provides the corresponding control function u as shown in Fig. 7.5. The solver generates the excitation in the form of rectangular oscillations. Of course, in a real

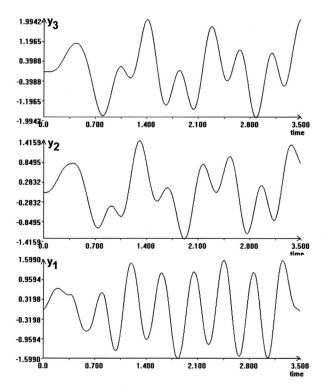

Figure 7.6 The form of the earthquake movement that results in the biggest amplitude of the structure movement.

earthquake we rather expect sinusoidal movements. Observe, however, that the model represents some kind of a low-pass filter. So, we can expect that the movements will be approximately the same if we replace $u(t)$ of Fig. 7.5 with a similar sinusoidal excitation.

The solver does not perform any frequency analysis. However, it detects the proper frequency or frequencies of the resonance and applies the control with these frequencies to maintain the trajectory at the boundary of the reachable set.

Fig. 7.6 depicts the movements of the three floors that correspond to the excitation u of Fig. 7.5. Note that although the system equations are stable, the oscillations do not disappear, and may even grow with time, while the ground movement amplitude does not.

7.4. Car suspension

Consider a car that moves forward and faces an irregularly shaped road. The car suspension receives a (nearly) vertical excitation from the point where the tire touches the ground. From the point of view of the driver, the car velocity is equal to zero and the road moves backwards. In our simplified model, the dynamics are considered only in the vertical direction, supposing that the suspension supports a "quarter car." In this model we have one wheel with a tire, spring, and shock absorber. Such model provides important information about the movements, although the 3D movement is ignored. If we use a complete system model, with four or more wheels, we must treat the car as a rigid body. This complicates the model, but may provide information about more realistic car movement and stability. The model used here is of order four, but it is not trivial, taking into account the included non-linearities.

Most of the models presented in the literature are given in the form of a system of four or six ODEs of the first order. Such models describe a "quarter car," that is, one wheel, spring, and shock absorber with only vertical movements with the mass of one quarter of the vehicle. The general properties and suspension types are described in "The Suspension Bible," available from the Web. To learn about more detailed models, we can turn to a lot of academic articles in the field. Allen et al. [1] discuss issues related to car stability, such as oversteer and high side-slip. The main topics are the lateral tire force and force saturation. The paper deals with the vehicle and tire characteristics and maneuvering conditions that cause the loss of directional control and potential tip-up and rollover.

Calvo et al. [2] describe some simulation experiments and compare the results with the real behavior of the damper. In the article, three mathematical models with increasing complexity are generated. The vehicle's behavior is analyzed for typical driving maneuvers taking into account lateral, vertical, and longitudinal forces. It is demonstrated that in order to obtain results with an acceptable level of accuracy, it is not necessary to rely on extremely complex shock absorber models.

The paper of Sayers et al. [12] shows how engineers can use simulation to rapidly gain experience in vehicle dynamics. The problems of steering, braking, and throttle inputs are demonstrated in an easy to use simulation environment. The paper describes features in the user interface that support quick learning on the part of the user. Van Kasteel et al. [13] deal with railway shock absorbers. Simulations of the absorbers are described based on comparison with measurements. This is used to help the manufacturers to tune the absorber parameters and to define the manufacturing requirements.

Chetan et al. [5] consider the fatigue analysis of the vehicle suspension system, using some analytic and experimental techniques. A review of the investigations is presented. In the paper of Nagarkar et al. [10], a model of order six is used for a quarter car from the suspension to the driver seat. The aim is to optimize the comfort and health criteria. Suspension spring stiffness, damping, and seat cushion stiffness are the design variables.

Simulations of a low-cost active suspension system of passenger vehicles are presented in the paper of Darling et al. [6]. A suspension with an active device responding to transducer signals on the vehicle is analyzed. This system aims to prevent roll and in this way improve passenger comfort and safety.

The dynamic reactions of wheels during breaking are investigated in the article of Peng and Hu [11]. They investigate the improvement one can expect from the implementation of different vehicle steering and driving mechanisms. Practical vehicle configurations such as four-wheel steering (4WS) and four-wheel drive (4WD) are considered. The optimization involves equality and inequality constraints. The problem is solved by non-linear programming techniques. Consult also CarBibles.com [3] and Magalhaes et al. [8].

A simplified model used here is a *DI model* that may provide more information about possible extreme system behavior, important in security issues.

Suppose that the car moves forward with a constant horizontal velocity V. We take the car's position as the reference and treat all variables as differences between their actual values and the initial equilibrium state. The variables are as follows:

y is the vertical movement of the wheel,

x is the vertical movement of the vehicle,

F_1 is the force determined by the compliance of the tire,

F_2 is the force of the suspension spring,

F_a is the force produced by the damper, and

u is an external excitation due to the shape of the road.

All the above variables are functions of time and all represent vertical movements and forces. Fig. 7.7 shows the corresponding mechanical scheme.

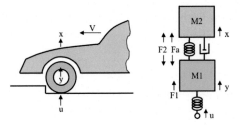

Figure 7.7 Car suspension scheme.

M_1 and M_2 are the mass of the wheel (together with the moving part of the suspension) and the mass of the car, respectively (M_2 being 1/4 of the mass of the chassis). To derive the model equations, we must use the force balance for the two masses, including the dynamic forces. This provides the following equations:

$$\begin{cases} F_1 - F_2 - F_a - M_1 \dfrac{d^2 y}{dt^2} = 0, \\ F_2 + F_a - M_2 \dfrac{d^2 x}{dt^2} = 0. \end{cases} \tag{7.6}$$

The forces of the two springs and of the damper depend on the corresponding changes of the spring and damper length. Thus,

$$\begin{cases} F_1 = K_1(y - u), \\ F_2 = K_2(x - y), \\ F_a = k_a(x' - y'). \end{cases} \tag{7.7}$$

Note that the damper force depends on the velocities and not positions. The above forces are supposed to be linear with respect to the positions and velocities. Further on, we will discuss a more realistic version that includes a non-linear damper. We convert the above equations to the canonical form of four equations of the first order, using the following notation:

$$x_1 = x, \quad x_2 = x', \quad x_3 = y, \quad x_4 = y'.$$

After substituting the new variables into (7.7) and reordering the equations, we obtain the following:

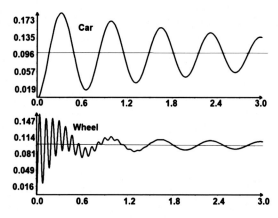

Figure 7.8 Car suspension simulation with the shock absorber coefficient equal to 200 N/(m/s).

$$\begin{cases} x_1' = x_2, \\ x_2' = [K_1(x_3 - x_1) + K_a(x_4 - x_2)]/M_2, \\ x_3' = x_4, \\ x_4' = [K_1(u(t) - x_3) - K_2(x_3 - x_1) - K_a(x_1 - x_2)]/M_1. \end{cases} \tag{7.8}$$

The last set of equations can be used to simulate our system. We obtain four equations, where the state vector of the model is $x = (x_1, x_2, x_3, x_4)$. This is correct because the movement equation for each mass is of the second order. This is a simplified model, whose purpose is to show an application of the DIs, rather than to obtain very exact quantitative data. We assume that the tire is always in contact with the road.

The model (7.8) is a simple linear version of the suspension movement. We start with this model to assess the appropriate value of the dumping coefficient K_a.

In the first simulation, the system parameters are as follows: $M_1 = 25$ kg, $M_2 = 200$ kg, $K_1 = 100,000$ N/cm, and $K_2 = 22,000$ N/cm.

The external excitation $u(t)$ is a step function with amplitude 10 cm, starting at $t = 0$ (the car hits a 10 cm high obstacle or sidewalk). Note that we use the SI unit system (mass in kg, velocity in m/s, and force in N). The first simulation was run with $K_a = 200$ N/(m/s). The position of the car body and the wheel can be seen in Fig. 7.8. The values are relative to the initial positions in the steady-state equilibrium.

The results with $K_a = 200$ N/(m/s) are not satisfactory. It is supposed that the driver should not detect more that three oscillations after entering

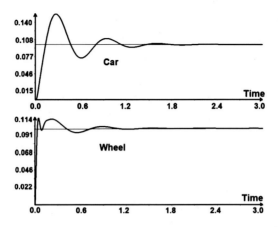

Figure 7.9 Simulation results with shock absorber coefficient equal to 1400 N/(m/s). The function $u(t)$ is constant, equal to 0.1 m.

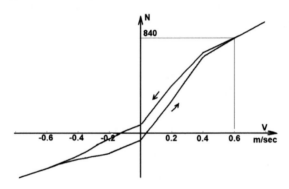

Figure 7.10 The velocity-force characteristics of the shock absorber.

an obstacle. Such a permissible case is reached with $K_a = 1400$, as shown in Fig. 7.9.

It can be seen that both the wheel and the car "jump" upwards by 10 cm. The wheel position oscillates with the dominant frequency higher than the car body. Anyway, both curves represent a superposition of two frequencies.

Now, we will use a more realistic model that includes a real shock absorber characteristic. The velocity–force plot is shown in Fig. 7.10. In a real shock absorber, this characteristic is non–linear. Moreover, the curve is charged with hysteresis. The characteristic with increasing force is different from the part of decreasing force. The velocity in Fig. 7.10 is negative for the compression movement. Such velocity sign is shown in Fig. 7.10

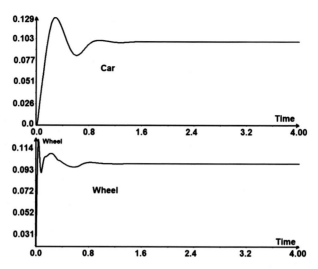

Figure 7.11 Simulation results with the shock absorber characteristics shown in Fig. 7.10.

in order to follow the commonly available shock absorber plots. In our equations, the velocity has opposite sign, i.e., it is positive for the compression movement. This means that when the car hits an obstacle (moving upwards), the force produced by the absorber is lower. This decreases the effect of the first upwards pulse. Fig. 7.11 shows a simple simulation result for this case.

Note that in Fig. 7.11, force and velocity are not related through Newton's equation of motion. This means that the force is not proportional to the acceleration. The velocity on the plot is the independent variable and the force is the result.

In this model, the hysteresis is modeled using the actual change of the speed v (acceleration). This is a simplified model, where the force jumps from one curve to another when the acceleration changes sign. In a real shock absorber this phenomenon is more complicated and there is an infinite number of curves on the velocity-force plot. This imperfection is the cause of some irregularities in the resulting plots of the car and wheel positions.

Now, consider a road with a certain configuration of obstacles. Suppose that the obstacle causes excitation $u(t)$ with an amplitude of ± 5 cm (Fig. 7.12). The problem is how big the range of car movements may be, supposing any, including the worst shape of the road.

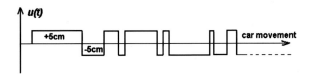

Figure 7.12 A possible shape of the road: the function $u(t)$ (vertical value exaggerated).

Let us treat u as an uncertain variable that can change within the given interval. Now, consider the right-hand side of (7.8) as a set obtained when u scans all the permissible values. In this way, (7.8) defines a *DI* because the right-hand side of the equation becomes a set-valued function. Using the DI solver, we can see the shape of the corresponding reachable set for the state variables. This may be useful when looking for unwanted extreme values that may influence the whole system and compromise the safety. As an example of an application of the DI solver we consider a suspension with non-linear shock absorber with no hysteresis. The hysteresis is omitted because the solver is based on some control theory methods that require certain continuity and regularity assumptions. The aim of the experiment is to learn how a malfunction of the absorber may cause big and dangerous car movements. Thus, we assume that the force produced by the bad absorber drops to only 15% of its normal value.

Let the model parameters be the same as in the simulation of Fig. 7.11, with 15% of the absorber force and simulation final time equal to 1.5 s. Fig. 7.13 shows the shape of the reachable set for time $= 1.5$.

In Fig. 7.13 the dots represent the end points of the trajectories. The horizontal and vertical coordinates correspond to the vertical movements of the car and wheel, respectively. Note that the chassis displacement may reach nearly 40 cm, while the excitement produced by the road shape is only ± 5 cm. This occurs because the DI solver detects the proper frequencies of the possible wheel and car oscillations and generates the excitements that make the model enter in resonance. Due to the non-linearity of the absorber, the trajectories are slightly displaced towards negative values of the car position.

Fig. 7.14 shows the projection of some of the model trajectories onto the time–chassis position plane. These are not random trajectories, but the trajectories that scan the boundary of the reachable set.

The solver also provides information about the road shape that corresponds to any point of the final reachable set. For example, Fig. 7.15 shows

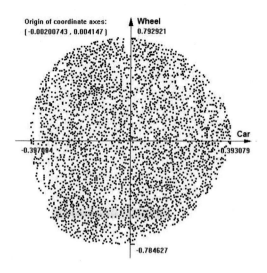

Figure 7.13 The shape of the reachable set at time = 1.5.

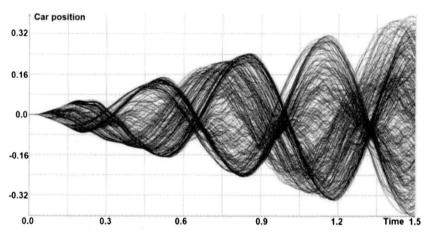

Figure 7.14 Projection of some boundary scanning trajectories on the time-car position plane.

the trajectories and the road shape that correspond to the maximal final positive displacement.

The program detects the two proper frequencies of the model and takes advantage of them to reach the boundary of the reachable set (in the case of Fig. 7.15, the greatest final car position). This is done without any spectral analysis.

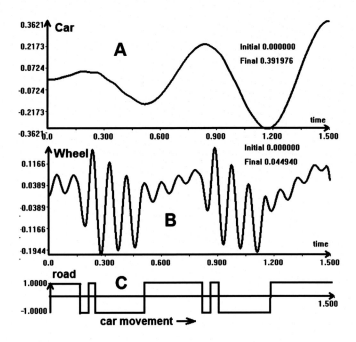

Figure 7.15 Chassis and wheel movement and the corresponding road shape.

The reachable set obtained in this way is several times greater than the intervals provided by conventional risk or sensitivity analysis. Treating the road displacement as the uncertain parameter, PowerSim-like analysis gives rather small variable intervals. Fig. 7.16 (left part) shows the results of conventional sensitivity analysis (small black section in the center) and the real shape of the possible final positions on the wheel-car plane at time equal to 1.5 s. This deficiency occurs because in conventional analysis the parameter remains constant along each model trajectory. It might appear that generating the road shape randomly within the permissible values (fluctuating in time), we could assess the shape of the reachable set. However, this is not true. The right-hand side of Fig. 7.16 shows the results of such simulation (10,000 trajectories) compared to the true reachable set.

Fig. 7.17 shows the trajectories corresponding to the experiments of Fig. 7.16, as functions of time.

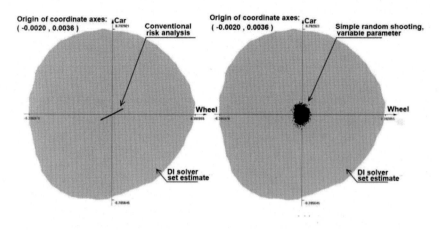

Figure 7.16 The reachable set for wheel and car final positions (gray area) and the results of conventional risk analysis and simple random shooting.

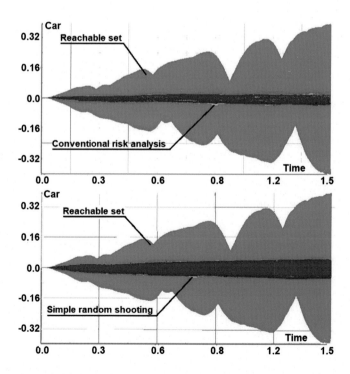

Figure 7.17 Reachable set, conventional risk analysis, and random shooting.

7.5. Conclusion

Using DIs, we can get a wide view on system dynamics. Simple simulation experiments normally provide sets of curves that show possible movements. Introducing uncertainty, we obtain attainable sets instead of single curves. This provides more relevant information. As stated before, this approach is deterministic. This is an alternative to methods that use stochastic models to deal with uncertain parameters. As for the example of the car suspension with weak shock absorber, the calculated reachable set shows how dangerous it is to drive on a bumpy road. The quarter-car simulation should be treated as a preliminary analysis. In further research, more complete results could be obtained while treating the vehicle as a rigid body with forces generated by four or more suspension systems. A significant difficulty is the visualization of the results provided by the DI solver. The reachable sets in such cases are given by multidimensional clouds of points, which are difficult to visualize in 2D projections.

References

[1] R. Allen, T. Rosenthal, D. Klyde, J. Hogue, Computer simulation analysis of light vehicles lateral/directional dynamic stability, Report, https://doi.org/10.4271/1999-01-0124, 1999.

[2] J.A. Calvo, B.L. Boada, J.L. San Roman, A. Gauchía, Influence of a shock absorber model on vehicle dynamic simulation, Proceedings of the Institution of Mechanical Engineers. Part D, Journal of Automobile Engineering 223 (2) (2009) 189–203, https://www.researchgate.net/publication/245391045.

[3] CarBibles.com, The suspension bible, Internet communication, http://www.carbibles.com/suspension_bible.html, 2016.

[4] N. Caterino, E. Cosenza, B.M. Azmoodeh, Approximate methods to evaluate Storey stiffness and interstory drift of RC buildings in seismic area, Structural Engineering and Mechanics 46 (2) (2013) 245–267, https://doi.org/10.12989/sem.2013.46.2.245.

[5] J. Chetan, P. Khushbu, M. Nauman, The fatigue analysis of a vehicle suspension system, International Journal of Advanced Computer Research (ISSN 2249-7277) 2 (4) (2012).

[6] J. Darling, R.E. Dorey, T.J. Ross-Martin, A low cost active anti-roll suspension for passenger cars, Journal of Dynamic Systems, Measurement, and Control 114 (4) (1992), https://doi.org/10.1115/1.2897730.

[7] J. Dimig, C. Shield, C. French, et al., Effective Force Testing: a Method of Seismic Simulation for Structural Testing, American Society of Civil Engineers, 1999, ISSN 0733-9445.

[8] H. Magalhaes, J.F.A. Madeira, J. Ambrósio, J. Pombo, Railway vehicle performance optimization using virtual homologation, Vehicle System Dynamics (ISSN 0042-3114) 54 (9) (2016), https://doi.org/10.1080/0042311042000266720.

[9] D. McCrum, M. Williams, An overview of seismic hybrid testing of engineering structures, Engineering Structures 118 (2016) 240–261.

[10] M.P. Nagarkar, G.J.V. Patil, R.N.Z. Patil, Optimization of nonlinear quarter car suspension–seat–driver model, Journal of Advanced Research (2016), https://doi.org/10.1016/j.jare.2016.04.003.

[11] H. Peng, J.S. Hu, Traction/braking force distribution for optimal longitudinal motion during curve following, Vehicle System Dynamics 26 (4) (1996) 301–320.
[12] M.W. Sayers, C.W. Mousseau, T. Gillespie, Using simulation to learn about vehicle dynamics, International Journal of Vehicle Design 29 (1–2) (2002) 112–127, https://doi.org/10.1504/IJVD.2002.002004.
[13] R. Van Kasteel, C.G. Wang, L. Qian, et al., A new shock absorber model for use in vehicle dynamics studies, SAE Transactions 43 (9) (2005) 613–631, https://doi.org/10.1080/0042311042000266720.

CHAPTER EIGHT

PID control: functional sensitivity

Abstract

In this chapter, the sensitivity of a control circuit with anti-windup PID controller is discussed. Instead of classical sensitivity, we use *functional sensitivity*, defined in terms of variational calculus. The non-local functional sensitivity is given in the form of the system's reachable set. The relation to differential inclusions is explained. The solution to the functional sensitivity is shown, using the differential inclusion solver, which calculates and displays the system's reachable sets. A comparison between functional sensitivity and the classical approach is made. The main simulation tool is the differential inclusion solver.

Keywords

Differential inclusion, Uncertainty, PID control, Reachable set

8.1. System sensitivity

By robust control systems we mean control circuits that function properly in environments where uncertain parameters and disturbances may influence the control outcome. Such circuits should not only be stable and, possibly, optimal, but they must also work under adverse circumstances. Here, the disturbances, as well as some system parameters are considered uncertain, but not necessarily random. We assume that such parameters may fluctuate in time, within certain limits. The robust control itself is not the main topic of the present chapter. We rather focus on the algorithm and software that calculates the reachable sets of dynamic systems. For more information on robust control, consult, for example, Xu and Lam [16] or Zhou and Doyle [17]. Let us make only some remarks and provide some references on system sensitivity analysis (SA).

Scatter plots represent a useful tool in SA. A plot of the output variable against individual input variables after (randomly) sampling the model over its input signals is made. This gives us a graphical view of the model sensitivity. View Friendly and Denis [7] for more details.

Regression analysis is a powerful tool for sensitivity problems. It allows us to examine the relationship between two or more variables of interest. The method is used to model the relationship between a response variable and one or more predictor variables or perturbations. Consult, for example, Freedman [6] or Cook [4].

Reachable Sets of Dynamic Systems
https://doi.org/10.1016/B978-0-44-313384-8.00007-5

For non–linear models, a useful tool is *variance-based modeling* or the Sobol method. It decomposes the variance of the output of the model or system into fractions which can be attributed to the input or sets of inputs. Consult Sobol [14].

The *screening method* decomposes the variance of the output of the model or system into fractions which can be attributed to model inputs. Thus, we can see which variable contributes significantly to the output uncertainty in high-dimensionality models. For more detail, see Morris [8].

The *logarithmic gain* is defined as the relative response of a dependent variable to a small changes in an independent variable. In dynamical systems, the logarithmic gain can vary with time, and this time-varying sensitivity is called dynamic logarithmic gain. This concept is used in dynamic SA, where the core model is a dynamic system, described by ordinary differential equations (ODEs). Consult Sriyudthsak et al. [15]. Consult also Section 1.5 for a discussion on classical and functional sensitivity.

System dynamics software offers tools for dynamic SA. Programs like Vensim or PowerSim include procedures that generate multiple model trajectories when the selected model parameters vary from one trajectory to another. However, in these packages the parameters are constant along the trajectory. Our approach is different. As explained in the following sections, we treat both the perturbations and uncertain model parameters as functions of time. Instead of the classical sensitivity concept, we use the functional sensitivity, defined in the following sections. The main tool used here is the differential inclusion (DI) solver, which calculates models' reachable sets (Raczynski [11]).

8.1.1 Functional sensitivity

Classical SA (the basic local version) is based on the partial derivatives of the model output Y, with respect to components of an input vector (model parameters) $u = (u_1, u_2, ..., u_n)$, at a given point u_0:

$$\left| \frac{\partial Y}{\partial u_i} \right|_{u_0}. \tag{8.1}$$

The derivative is taken at some fixed point in the space of the input variables (hence the "local" in the name of the analysis mode). The use of partial derivatives suggests that we consider small perturbations of the input vector, around the point of interest u_0. Consult Cacuci [3].

Consider a dynamic model described by an ODE,

$$\frac{dx}{dt} = f(x, u, t), \tag{8.2}$$

where $x = (x_1, x_2, ..., x_n)$ is the state vector, $u = (u_1, u_2, ..., u_m)$ is the perturbation (parameters, control) vector, t is the time, $x \in X$, $u \in U$, and $f : X \times U \times R \rightarrow X$. Here, X is the state space, U is the control space, and R is the real number space. We restrict ourselves to the case $X = R^n$, $U = R^m$, $R = R^1$, R^k being the real Euclidean k-dimensional space. Let $t \in I = [0, T]$, and let G be the space of all measurable functions $u : I \rightarrow R^m$.

Now, consider a variation δu of u and a perturbed control u_p. The variation is a function of time, so that $u_p(t) = u(t) + \delta u(t)$ $\forall t \in I$. The solution to (8.2) over I with given initial condition $x = x_0$ and given function $u_p(t)$ will be called a *trajectory* of (8.2). Thus, any component x_k of the final value of $x(t)$ depends on the shape of the whole function u_p. In other words, $x_k(t) = x_k(t)[u_p]$ (for any fixed t) is a functional (not a function) of u_p. Unlike a function, in our case, the functional x_k is a mapping from the space G to R. Denote $\delta x_k = x_k[u + \delta u] - x_k[u] = x_k[u_p] - x_k[u]$.

In this book, the *local functional sensitivity* is defined as

$$S_k = \left| \frac{\delta x_k}{\delta u_0} \right|. \tag{8.3}$$

Note the difference between the conventional local sensitivity (8.1) and the functional sensitivity (8.3). The notation δu is nothing new; it denotes the *variation* of the function u, as defined in the calculus of variations (Nearing [9] and Elsgolc [5]). However, the sensitivity defined in (8.3) is quite different from the classical definition (8.1).

The term S_k of (8.3) defines a local property of the trajectory $x_k(t)$. Here, we are interested in the response of the model to perturbations that are not necessarily small. We will use the (non-local) *functional sensitivity*, defined as the *set of the graphs of all trajectories* of (8.2), where $u = u_0 + \Delta u$. Here, $\Delta u(t)$ is a limited perturbation, not necessarily small. This is equivalent to saying that $u(t)$ belongs to a set of restrictions $C(t)$, $u(t) \in C(t)$ $\forall t \in I$. Here, $C(t) \subset R^m$ is the restriction set for the control. The functional sensitivity defined this way is non-local. We do not use the term "global," because this is not a global property of the model. We only do not require the perturbation to be small.

Models with uncertain parameters and control systems are closely related to DIs. Consider a model defined as follows:

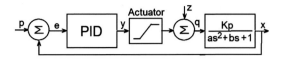

Figure 8.1 Feedback control with disturbance.

$$\frac{dx}{dt} = f(x(t), u(t), t), \quad x(t_0) \in R^n, \quad u(t) \in C(x, t), \quad t \in I = [0, T]. \quad (8.4)$$

Eq. (8.4) may represent a control system with control variable u, as well as a dynamic model with uncertain variable parameters u. When u scans all possible values in C, the right-hand side of (8.4) scans the values inside a set F. This defines the corresponding DI as follows:

$$\frac{dx}{dt} \in F(x, t), \quad (8.5)$$

where $F(x, t) = \{z : z = f(x, u, t), u \in C(x, t)\}$. This way, we obtain a DI. More detailed assumptions and a comprehensive survey on DIs can be found in Aubin and Cellina [2].

8.2. Proportional, derivative, and integral control action

Fig. 8.1 shows a closed-loop control system. The controller is of the proportional–derivative–integral action (PID) type, and the process is of second order and oscillatory. The set point is denoted as p, e is the control error, z is a disturbance, and q and x are the process input and output, respectively. The control error e is equal to $p - x$.

At the controller output there is an actuator that may saturate.

The classical PID controller is described by the following equation (in terms of Laplace transform):

$$y(s) = K_r e(s) \left(1 + T_d s + \frac{1}{T_i s}\right), \quad (8.6)$$

where y is the controller output, e is the control error (controller input), K_r is the controller gain, T_i is the integrator parameter (duplication time), and T_d is the derivative action parameter.

To avoid differentiation of the set point signal, which may have discontinuities, we use the controller where the set point is not applied to the differentiator (a special case of "set point weighing"). In the time domain,

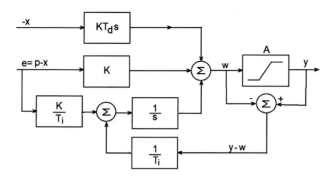

Figure 8.2 Anti-windup PID: analog version.

the controller equation is as follows (the prime mark means time differentiation):

$$y(t) = K_r \left(e(t) + T_d x'(t) + \frac{1}{T_i} \int_0^T e(t)\,dt \right). \qquad (8.7)$$

Recall that the derivative action in (8.7) is used to make the controller response faster in case of rapidly changing perturbations. The integral term reduces the static control error in a somewhat slower way.

8.2.1 Windup in PID controllers

One of the problems in PID control is the *windup* effect. This occurs when the control error is different from zero for a relatively long time interval. In this case, the value of the integral term unnecessarily grows and may result in bad and long transient processes. In this chapter, two versions of the controller are used: the classical PID controller and the *anti-windup* PID controller.

There are several analog versions of the anti-windup controller (consult Åström and Hägglund [1]). A possible implementation is shown in Fig. 8.2.

This is an analog PID anti-windup scheme, where a feedback is added that turns off the integrator when the actuator at the output of the controller reaches the saturation level. The actuator may be a real, physical actuator or a model of the actuator. If the actuator operates within the linear region, then the signal $y - w$ is equal to zero, and we have a conventional PID action. When the saturation level is reached, then the non–zero $y - w$ signal appears. It makes the input to the integrator equal to zero, and the integrated value no longer grows.

In digital versions of anti-windup PID, windup is circumvented by soft-ware, like in the controller we use in the following.

Let us denote the value generated by the integrator as $g(t)$:

$$g(t) = \int_0^t e(\tau)\,d\tau.$$

Consider a controlled process described by the following equation:

$$ax''(t) + bx'(t) + x(t) = K_p q(t), \tag{8.8}$$

where $q(t)$ is the input signal and $x(t)$ is the process response.

Now, denote

$$x_1 = x, \quad x_2 = x', \quad x_3 = g.$$

The state equations for the control circuit of Fig. 8.1 are as follows:

$$\begin{cases} x_1' = x_2, & \\ x_2' = \dfrac{1}{q}(K_p(y+z) - bx_2 - x_1), & \text{process} \\ x_3' = \dfrac{1}{T_i}(p - x_1), & \text{integrator} \\ y = K_r(e - T_d x_2 + x_3), & \text{controller output} \\ e = p - x_1. & \end{cases} \tag{8.9}$$

When calculating the controller output, we do not need any special treatment for the derivative action. This value is taken from the equations of the controlled process, state variable x_2. The derivative is calculated for the process output only.

The variable v is added to simulate the anti-windup PID controller. This PID version is implemented to avoid the windup effect when the value of the integration action may become too big. This occurs when the control error is positive or negative during long time intervals. The variable v is defined as follows:

$$v(e, y) = \begin{cases} 1 & \text{if } |y| < L, \\ 1 & \text{if } y \geq L \text{ and } e < 0, \\ 1 & \text{if } y \leq -L \text{ and } e > 0, \\ 0 & \text{otherwise.} \end{cases} \tag{8.10}$$

Here, L is the saturation level of the actuator. This variable shuts down the integrator when the actuator reaches its saturation level (see Eq. (8.9)).

Before showing the simulation results and reachable sets, let us define some new concepts.

8.3. Differential inclusion solver

Someone working with differential equations can find a huge number of numerical methods and software programs. On the other hand, for DIs there is nearly nothing that could help a simulationist. The DI solver is the result of an attempt to fill this gap in simulation software. The basic version of the DI solver is not new; it was published in 2002 [11]. Here, we only recall the algorithm. In this chapter, a new version of the solver and some new applications to functional sensitivity are mentioned. It should be emphasized that to obtain the reachable set we do not apply simple deviations of the parameters. The solver algorithm discussed here is rather complicated and based on the Jacobi–Hamilton equations and some results from optimal control theory.

The solver needs the Embarcadero Delphi package installed on the user machine. A limited stand-alone ".exe" version of the solver is also available. It should be noted that our main goal is the determination of the reachable set and not optimization. The DI solver and the present problem statement should not be confused with the DI method used in optimal control problems.

To avoid repetitions from earlier chapters, we will not discuss here the algorithm of the DI solver in detail (see Raczynski [11–13] and Chapter 2 of this book). In few words, the DI solver generates a series of DI trajectories that scan the boundary, and not the interior, of the reachable set. It would be an error to look for a uniformly distributed cloud of points in the interior of the reachable set. One could suppose that to assess the shape of the reachable set, we can generate a number of random trajectories that belong to its interior, and see the boundary. However, this is not true. Such simple random shooting gives incomplete results that do not resemble the true shape of the reachable set.

The solver algorithm uses some results from optimal control theory (Pontryagin [10]). From Pontryagins' maximum principle it is known that each model trajectory that reaches a point on the boundary of the reachable set at the final simulation time must entirely belong to the boundary of this set for all earlier time instants. Moreover, such trajectory must satisfy the Jacobi–Hamilton equations (a necessary condition). These equations involve a vector of auxiliary variables $p = (p_1, p_2, p_3, ..., p_n)$.

In optimal control, the problem is that we have the initial conditions for the state vector x, but the conditions for the vector p are given at the end of the trajectory ($t = T$). If no analytic solution is available, we must use an iterative procedure for the so-called *two-point boundary value* (TPBV) problem. In our case, we are in a better situation. Observe that starting with the given initial condition for x, and with any initial condition for p, we obtain a trajectory that belongs to the boundary of the reachable set. This means that we do not have to solve the TPBV problem. The algorithm generates a series of trajectories with randomly generated initial conditions $p(0)$. After integrating a sufficient number of trajectories, we can see the shape of the boundary of the reachable set. In other words, we are looking for the mapping $MP : R^n$ (space of initial conditions for p) $\rightarrow RS(T)$, where $RS(T)$ is the boundary of the reachable set at $t = T$. The problem is that the mapping MP may be extremely irregular even for a simple linear model. The solver algorithm uses a certain heuristic procedure to avoid "holes" in the final image (Raczynski [11]). The new version of the solver has been developed for multiprocessor machines. Considerable acceleration has been achieved by calculating the model trajectories concurrently.

8.4. Calculating reachable sets

As stated before, the functional sensitivity is given by the reachable set of the system model. For the reader's convenience, let us repeat here the equations of the control system under consideration:

$$\begin{cases} x_1' = x_2, \\ x_2' = \dfrac{1}{q_v}(K_p(y + z) - bx_2 - x_1), & \text{process} \\ x_3' = \dfrac{q_v}{T_i}(p - x_1), & \text{the integrator} \\ y = K_r(e - T_d x_2 + x_3), & \text{controller output} \\ e = p - x_1, \end{cases} \qquad (8.11)$$

where the factor v is defined by (8.10).

The model parameters are as follows:

$a = 2$, $b = 1$, $K_p = 1$ (process parameters)
$p = 1$ (the set point)
$K_r = 4$ (controller gain)
$T_i = 2$ (duplication time)
$T_d = 0.2$ (derivative time)

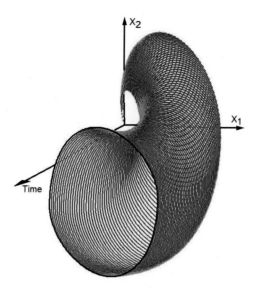

Figure 8.3 Open-loop reachable set for the process only. 3D image generated by the DI solver.

The above PID settings are not optimal. They were defined this way to exaggerate some imperfections in the transient process. Anyway, a robust control system should work not only with optimal parameters. Our model contains eight parameters that may have uncertain values. Using the DI solver, we can calculate reachable sets related to any uncertain parameter and for any combination of them. This can result in many simulations and generated images. Here, let us show only some results and images to illustrate the possible experiments.

Suppose that both the values of the disturbance z and the process gain K_p are uncertain and may fluctuate by $\pm20\%$ around their original values. Without perturbation, the system response reaches a steady state after about 30 time units. However, we are interested rather in system behavior during the transient process, so we will investigate the functional sensitivity at a shorter time, $t = 8$. Fig. 8.3 shows the reachable set of the process response to unit step input, for the process only, with open loop. In Fig. 8.4 we can see some randomly selected (but not random) trajectories that scan the boundary of the reachable set for the process only.

Now, let the same parameter fluctuations be applied in the closed-loop control circuit. First, we simulate the system without actuator saturation. The closed-loop model is of order three. Thus, the boundary of the reachable set at the given final time instant is not just a contour, but a 3D cloud

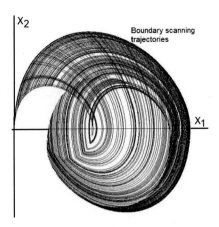

Figure 8.4 Some boundary scanning trajectories, process in open loop.

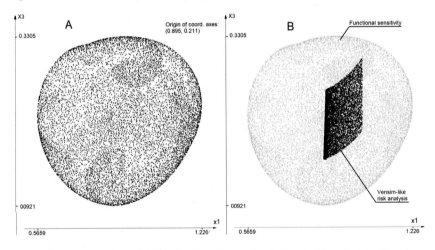

Figure 8.5 Functional sensitivity of a closed-loop circuit. Reachable set at $t = 8$, with no saturation.

of points. The points are located on a 3D surface, like a surface of a balloon. Consequently, we cannot get an image like that of Fig. 8.3. We can only see projections of the boundary of the reachable set onto a 2D plane. The DI solver generates a lot of such projections. Let us show only some examples, namely, the projections onto the planes (x_1, x_3) and (time, x_1). Fig. 8.5 shows the projection of the reachable set at $t = 8$ onto the plane (x_1, x_3).

Fig. 8.5A shows the end point of the trajectories. All these points lie on the boundary of the reachable set; what we see is a 2D projection of the 3D

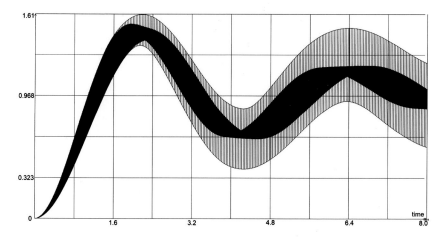

Figure 8.6 Projection of the reachable set onto the time-x_1 plane.

surface. In Fig. 8.5B we can see a comparison of the functional sensitivity with the Vensim-like analysis, where the uncertain parameters change from one trajectory to another, but remain constant along the trajectory (the same, ±20% changes).

Fig. 8.6 depicts the "side view" of the same reachable set, projected onto the time-x_1 plane.

Now, consider the anti-windup controller with actuator with saturation level equal to 1.6. Fig. 8.7 shows, like Fig. 8.5, the projection of the boundary points at the $x_1 x_3$-plane. Panel B shows also the points obtained by classical risk analysis, with the same range of parameter changes. The difference in the shape of the reachable set can be observed. Note the difference between the maximal value of x_1 (overshoot) in Figs. 8.5 and 8.7, equal to 1.61 and 1.35, respectively. As can be seen in Fig. 8.6, the functional sensitivity coincides with the classical analysis for the initial interval (2 time units), but after this interval the results are very different. The functional sensitivity set is several times greater than that provided by the classical method. In general, this occurs when the model contains oscillatory components. See Fig. 8.8.

8.5. Conclusion

Functional sensitivity is closely related to DIs. Here, we use the non-local functional sensitivity, which is defined as the reachable set of the model with uncertain parameters. This approach is deterministic. We do

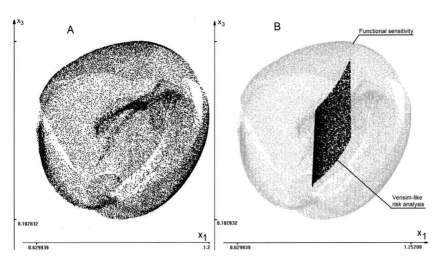

Figure 8.7 Functional sensitivity of a closed-loop circuit. Reachable set at $t = 8$, with anti-windup controller, with saturation level 1.6.

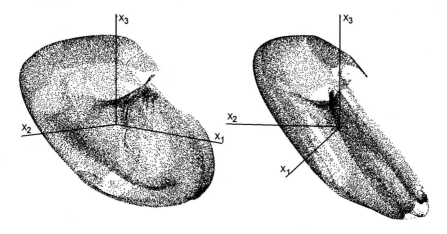

Figure 8.8 The reachable set viewed from different angles, with saturation level equal to 1.6.

not treat the disturbances as random variables. Such deterministic approach seems to be more useful in problems of robust control, when we are interested in "worst-case" behavior rather than in probabilistic model properties.

The DI solver works satisfactorily for the presented model. Classical risk analysis may, in some cases, coincide with the functional sensitivity. However, if the model includes oscillatory components, the results may be very different, and the functional sensitivity sets may be several times greater than those obtained by classical methods. An essential characteristic of the DI solver is that it explores the boundary of the reachable set and not its interior. Though some mechanisms of Pontryagin's maximum principle are used, we do not solve any *two-point boundary value* problem. To estimate the shape of the reachable set for a second-order model we need between 30 seconds and 2 minutes. If the non-linearities are strong, more time may be required. The calculation time increases significantly for higher-order models. A strongly non-linear model of order six may require more than 30 minutes of computing time.

Some imperfections of the obtained images may appear when the reachable set is complicated and folds several times. This is normal; recall that the maximum principle provides necessary and not sufficient conditions for optimal control. Another cause of imperfections may be the model non-linearity. The Jacobi–Hamilton equations require the model functions to be twice continuously differentiable, which is not always the case. However, in such cases the solver keeps working. Anyway, all the points provided by the solver belong to the reachable set, so we always obtain the estimation "from below." Further research should be done to accelerate the algorithm and improve the graphical display of results for multidimensional models.

References

[1] K.J. Astrom, T. Hägglund, PID Controllers, Instrument Society of America, ISBN 1-55617-516-7, 1995.

[2] J.P. Aubin, A. Cellina, Differential Inclusions, Springer-Verlag, Berlin, ISBN 978-3-642-69514-8, 1984.

[3] D.G. Cacuci, Sensitivity and Uncertainty Analysis, Chapman & Hall/CRC, London, ISBN 1-58488-115-1, 2003.

[4] R.D. Cook, S. Weisberg, Criticism and influence analysis in regression, Sociological Methodology 13 (1982) 313–461.

[5] L.D. Elsgolc, Calculus of Variations, ISBN 978-0486457994, 2007.

[6] D.A. Freedman, Statistical Models: Theory and Practice, Cambridge University Press, 2005.

[7] M. Friendly, D. Dennis, The early origins and development of the scatterplot, Journal of the History of the Behavioral Sciences 41 (2) (2005) 103–130, https://doi.org/10.1002/jhbs.20078.

[8] M.D. Morris, Factorial sampling plans for preliminary computational experiments, Technometrics 33 (1991) 161–174, https://doi.org/10.2307/1269043.

[9] J. Nearing, Mathematical Tools for Physics, Dover Publications, 2010.

[10] L.S. Pontryagin, The Mathematical Theory of Optimal Processes, Wiley Interscience, New York, 1962.

[11] S. Raczynski, Differential inclusion solver, in: Conference Paper: International Conference on Grand Challenges for Modeling and Simulation, The Society for Modeling and Simulation Int., San Antonio, TX, 2002.

[12] S. Raczynski, Differential inclusion approach to the stock market dynamics and uncertainty, Economy Computing Optimization Risks Finance Administration Net Business (ISSN 2444-3204) 1 (1) (2015) 58–65.

[13] S. Raczynski, Uncertainty treatment in prey-predator models using differential inclusions, Nonlinear Dynamics, Psychology, and Life Sciences (ISSN 1090-0578) 22 (4) (2018) 421–438.

[14] I. Sobol, Sensitivity analysis for non-linear mathematical models, Mathematical Modeling and Computational Experiment 1 (1993) 407–414.

[15] K. Sriyudthsak, H. Uno, R. Gunawan, F. Shiraishi, Using dynamic sensitivities to characterize metabolic reaction systems, Mathematical Biosciences 269 (2015) 153–163.

[16] S. Xu, J. Lam, Robust Control and Filtering of Singular Systems, Springer, 2006.

[17] K. Zhou, J.C. Doyle, Essentials of Robust Control, Prentice Hall, ISBN 0-13-525833-2, 1997.

CHAPTER NINE

Speed control of an induction motor

Abstract

A new approach to system sensitivity is applied to the constant voltage-to-frequency ratio (V/f) speed control of induction motors. Unlike the classical sensitivity definition, here, the *functional sensitivity* is defined in terms of variational calculus. The modeling tool used here is the *differential inclusion*. Some basic definitions are recalled, including the *reachable set* of the differential inclusion. The *functional sensitivity* of the V/f speed control is proposed, based on the reachable sets of differential inclusions.

It is pointed out that the non-local functional sensitivity is given in the form of the system's reachable set. The solution to the functional sensitivity is shown, using the differential inclusion solver, which calculates and displays the system's reachable set. A new differential inclusion solver is used to generate the reachable and sensitivity sets of a model of the V/f speed control circuit. A comparison between functional sensitivity and the classical approach is made.

Keywords

Induction motor, Simulation, Sensitivity, Reachable set, Speed control

Nomenclature

T torque (Nm)
f_n nominal frequency
V supply voltage/1.73
z_1 stator impedance
r_2 rotor resistance
n_s synchronous velocity (rpm) (supposed to be equal to 1800 for $f = 60$ Hz in the following)
s $(n_s - n)/n_s$ (the slip)
n motor velocity (rpm)
P number of pole pairs (supposed to be equal to 4 in the following)

9.1. Introduction: model sensitivity

In this chapter we discuss as an example the reachable sets of a speed control circuit for an induction motor. This model has been selected because of the strong non-linearities. The reachable sets can be used in designing a robust control system, based on the knowledge about the model sensitivity.

Reachable Sets of Dynamic Systems
https://doi.org/10.1016/B978-0-44-313384-8.00008-7

In the following sections, a short overview of system sensitivity is provided and the problem of induction motor speed control is stated. Then, the concept of *functional sensitivity* (see Section 1.5, Chapter 1) is recalled. Finally, the *differential inclusion (DI) solver* is described, and an application to V/f speed control is shown.

An overview of the methods of *sensitivity analysis* (SA) can be found in Section 1.5. To avoid repetition, we will not comment on the conventional SA methods. For a short overview of *scatter plots, regression analysis*, the *Sobol method*, logarithmic gain, and *system dynamics SA*, consult Section 1.5. See also [1–6].

Our approach to SA is different. As explained in the following sections, here, both the perturbations and uncertain model parameters are treated as functions of time. Instead of the classical sensitivity concept, *functional sensitivity* is used, defined in Section 1.5. The main tool used here is the DI solver, which calculates models' reachable sets (Raczynski [7]).

In functional sensitivity, the disturbances, as well as changing model parameters, are functions of time. The problem statement is based on the variational calculus approach, and not on partial derivatives as in classical SA. The main tool used here is the *DI* instead of the ordinary differential equation (ODE). The reachable sets of DIs are calculated using the new DI solver. The images of the sensitivity sets provided by the solver can help in robust control design. Such images can hardly be found in the literature. This kind of analysis may be useful in the design of robust control systems (consult Zhou [9]).

9.2. V/f speed control of an induction motor

The angular velocity w of a motor is given by the following equation (the prime sign stands for time differentiation):

$$x' = \frac{T - L}{I}, \tag{9.1}$$

where T is the torque (Nm), L is the mechanical load (Nm), and I is the moment of inertia of the rotor (kg m^2). The formula for T used in this chapter is as follows:

$$T(V, s) = \frac{k V^2 r}{\left(z_1 + \dfrac{r_2}{s}\right)^2 s n_s}, \quad k = \frac{f_n}{2\pi}, \tag{9.2}$$

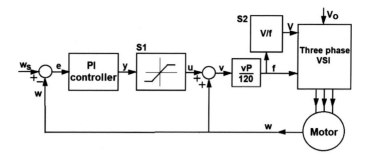

Figure 9.1 Block diagram of V/f speed control.

where T is the torque (Nm), f_n is the nominal frequency, V is the supply voltage/1.73, z_1 is the stator impedance, r_2 is the rotor resistance, n_s is the synchronous velocity (rpm) (supposed to be equal to 1800 for f = 60 Hz in the following), $s = (n_s - n)/n_s$ (the slip), n is the motor velocity (rpm), and P is the number of pole pairs (supposed to be equal to 4 in the following).

Unlike the DC motor, the velocity of the induction motor cannot be easily controlled by changing the supply voltage or stator current. Three-phase voltage source inverters (VSIs) must be used. These devices are based on insulated-gate bipolar transistor (IGBT) semiconductor switches. There are alternatives to the IGBT: insulated-gate commutated thyristors (IGCTs) and injection-enhanced gate transistors (IEGTs). Here, VSI technology is not discussed. An ideal VSI is assumed that provides the three-phase supply of voltage V and frequency f, receiving a necessary power supply and the desired values of V and f. For more information about VSI control, consult Mujal-Rosas and Orrit-Prat [10] or Chen et al. [11].

Mathematical modeling of induction motors is discussed in the article of Conrad et al. [12]. A similar discussion on induction motor modeling can be found in the articles [13] and [14].

The velocity of the motor can be changed by changing the frequency. However, such control with constant voltage makes the stator current grow for low frequencies, producing saturation of the air gap flux. Therefore, the stator voltage should be reduced according to the frequency to maintain the air gap flux constant. The magnitude of the stator flux is (approximately) proportional to the ratio of the stator voltage and the frequency. Hence, if the voltage-to-frequency (V/f) ratio is kept constant, the flux remains constant. This method is referred to as V/f speed control. Fig. 9.1 shows a typical scheme of V/f control, with a PI controller.

Let w_s be the set point (desired velocity, rpm). We have

$$
\begin{cases}
e = w_s - w, & \text{(control error)} & \text{(A)} \\
y = K\left(e + \frac{z}{T_i}\right), & & \text{(B)} \\
z(t) = \int_0^t e(\tau)d\tau, & & \text{(C)} \\
v = u + w, & & \text{(D)} \\
f = \dfrac{vP}{120}, & & \text{(E)} \\
V = V_{ref}\dfrac{f}{60}, & & \text{(F)} \\
\dfrac{dw}{dt} = \dfrac{T(V,s) - L}{I}, & & \text{(G)}
\end{cases}
\qquad (9.3)
$$

where

$$
u = \begin{cases}
y & \text{if } |y| \le y_m, \\
-y_m & \text{if } y < -y_m, \\
y_m & \text{if } y > y_m.
\end{cases}
\qquad (9.4)
$$

Here, I is the moment of inertia of the rotor (kg m^2) and L is the load (Nm).

It is assumed that $V_{ref} = 440/\sqrt{3} = 254.03$. The velocities w_s, w, v, and u are given in rpm units, and frequency f is given in Hz. From Eq. (9.3C) we have

$$
\frac{dz}{dt} = e(t).
\qquad (9.5)
$$

Eqs. (9.3G) and (9.5) form the ODE model of the system dynamics, where w and z are the state variables. Block S1 of Fig. 9.1 is a delimiter. Block S2 also includes the necessary saturation restrictions.

To illustrate the action of the PI controller, let us show a simple simulation of the control circuit of Fig. 9.1. The model parameters are as follows: $K = 0.5$, $T_i = 2$ s, $y_m = 200$, $P = 4$, $V_{ref} = 254$ V, $L = 10.1$ Nm, and $I = 0.08$ kg m^2.

The simulation starts with the motor rotating at 1800 rpm (synchronous speed, set point), with no load. The load is applied at time $= 0$. First, the velocity decreases because of the load. The PI controller makes the velocity go back to 1800 rpm. The presence of the integral part of the controller makes the steady-state velocity exactly equal to the set point. At time $= 50$, the set point changes to 1900 rpm. The transient process can be seen in Fig. 9.2. It should be noted that in this simulation and in the following SA,

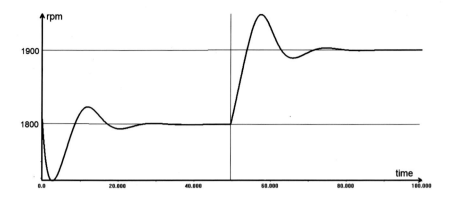

Figure 9.2 Simple simulation of the control system of Fig. 9.1.

the controller settings are not optimal. The duplication time T_i is intentionally set to a relatively small value to make the transient processes oscillatory. The controller gain is rather small to avoid saturation and instability.

9.3. Functional sensitivity

9.3.1 Differential inclusions

Let us recall some concepts of *DIs*. A more detailed overview can be found in Chapter 1 of this book.

Consider a model defined as

$$\begin{cases} \dfrac{dx}{dt} = f(t, x(t), u(t)), \quad x(t) \in R^n, \\ u(t) \in C(x, t), \quad t \in I = [0, T], \quad T > 0, \end{cases} \tag{9.6}$$

where R^n is the real n-dimensional Euclidean space, $x \in R^n$, $u \in R^m$, t is a real variable representing time, and C represents the restrictions for variable u. Eq. (9.6) may be interpreted as a control system with control variable u, or as a model with uncertain parameter u.

When u scans all possible values in C, the right-hand side of (9.6) scans the values inside a set F. This way, the corresponding DI is obtained as follows:

$$\frac{dx}{dt} \in F(t, x). \tag{9.7}$$

Here, $F(t, x) = \{z : z = f(t, x, u), u \in C(x, t)\}$. More detailed assumptions and a comprehensive survey on DIs can be found in Aubin and Cellina [15].

9.3.2 Local and non-local sensitivity

For the reader's convenience, let us repeat here some notions of *functional sensitivity*. A more detailed presentation can be found in Chapter 1.

Most of the known methods of SA use the partial derivatives of the model output Y, with respect to components of the model parameters $u = (u_1, u_2, ..., u_n)$, at a given point u_0:

$$\left| \frac{\partial Y}{\partial u_i} \right|_{u_0}. \tag{9.8}$$

This derivative is taken at some fixed point in the space of the input. The use of partial derivatives suggests that small perturbations of the input vector are considered, around the point of interest u_0. Consult Cacuci [16].

Consider a dynamic model described by an ODE,

$$\frac{dx}{dt} = f(x, u, t), \tag{9.9}$$

where $x = (x_1, x_2, ..., x_n)$ is the state vector, $u = (u_1, u_2, ..., u_m)$ is the perturbation (varying parameters, control) vector, and t is the time. Let also $x \in X$, $u \in U$, $f : X \times U \times R \to X$. X is the state space, U is the control space, and R is the real number space. Here, $X = R^n$, $U = R^m$, $R = R^1$, R^k being the real Euclidean k-dimensional space. Let $t \in I = [0, T]$, and let G be the space of all measurable functions $u : I \to R^m$.

By δu we mean a variation of u. The variation is time-dependent. Denote $u^*(t) = u(t) + \delta u(t) \ \forall t \in I$. The solution to (9.9) over I with given initial condition $x = x_0$ and given function $u(t)$ will be called a trajectory of (9.9). Any component x_k of the final value of $x(t)$ depends on the shape of the whole function u. In other words, $x_k(t) = x_k(t)[u^*]$ is a functional (not a function) of u^*. Unlike a function, in our case, the functional is a mapping from the space G to R. Denote $\delta x_k = x_k[u + \delta u] - x_k[u] = x_k[u^*] - x_k[u]$.

Let us define the *local functional sensitivity* as follows:

$$S_k = \left| \frac{\delta x_k}{\delta u_0} \right|. \tag{9.10}$$

The notion of functional sensitivity is quite different from the conventional local sensitivity (9.8). The notation δu denotes the variation of the function u, as defined in the calculus of variations (Nearing [17] and Elsgolc [18]). A variational approach to sensitivity is also discussed in Arora [19], Mordukhovich [20], and Sriyudthsak [5].

Eq. (9.10) defines a local property of the trajectory $x_k(t)$. However, the analyst is frequently interested rather in the response to perturbations, which are not necessarily small. Our task is to define the functional sensitivity as the set of the graphs of all trajectories of (9.9), where $u = u_0 + \Delta u$. Here, $\Delta u(t)$ is a limited perturbation, not necessarily small. For the control system (9.10), it is equivalent to saying that $u(t)$ belongs to a set of restrictions $C(x, t)$, $u(t) \in C(x, t)$, $\forall t \in I$. Here, $C(x, t)$ is a subset of R^m. When u scans all possible values inside the set C, the right-hand side of (9.9) becomes a set-valued function. This way, (9.10) with disturbed control also defines the corresponding DI, that defines the *non-local functional sensitivity*. The term "global" is not used here because this is not a global property of the model. The perturbation is just not required to be small.

9.3.3 Differential inclusion solver

The DI solver is not new. It was published in Raczynski [7]. See also Chapter 2 of the present book. Let us recall some properties of the solver.

The DI solver generates a series of DI trajectories that scan the boundary, and not the interior, of the reachable set. It would be an error to look for a uniformly distributed cloud of points in the interior of the reachable set. What is needed is the boundary of the set, which can be defined by a smaller number of attainable points. One could suppose that to assess the shape of the reachable set, a number of trajectories that belong to its interior can be generated, to see the boundary. However, this is not true. Such simple random shooting gives incomplete results that do not resemble the true shape of the reachable set.

In few words, the solver algorithm uses some results from optimal control theory (Pontryagin [8]). From the Pontryagin maximum principle it is known that each model trajectory that reaches a point on the boundary of the reachable set at the final simulation time must entirely belong to the boundary of this set for all earlier time instants. Such trajectory must satisfy the Jacobi–Hamilton equations (a necessary condition). These equations involve a vector of auxiliary variables $p = (p_1, p_2, p_3, ..., p_n)$.

The algorithm generates a series of trajectories with randomly generated initial conditions $p(0)$. After integrating a sufficient number of trajectories, the shape of the boundary of the reachable set can be seen. Consult Raczynski [7].

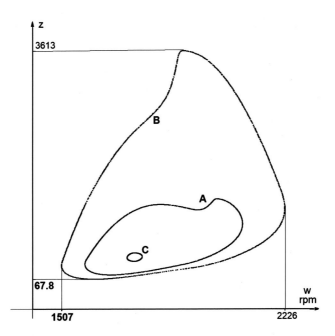

Figure 9.3 System reachable sets for time = 20.

9.4. Functional sensitivity of V/f control systems

Now, let us have a look at the functional sensitivity of our model. Here, the focus is on the functional sensitivity and not on the model itself. Such analysis can be useful when dealing with robust control design. Unlike conventional SA, functional SA is dynamic, based on DIs. Note that the reachable sets calculated here are not obtained by an application of simple disturbances. We use the DI solver described in Chapter 2.

The control system of Fig. 9.1 has two external signals that may be treated as disturbances: the load L and the set point w_s. Of course, any other model parameter can be changed. If the variables L and w_s scan the interior of a given permissible set, then the right-hand sides of (9.3G) and (9.5) scan a certain set $F \subset R^2$. This is the right-hand side of the corresponding DI. The reachable set of this inclusion defines the functional sensitivity of our model.

Fig. 9.3 shows the boundaries of the reachable sets with different model parameters. These are intersections of the reachable sets with the plane time = 20. Model parameters are as follows: $K = 0.5$, $T_i = 2$ s, $y_m = 200$,

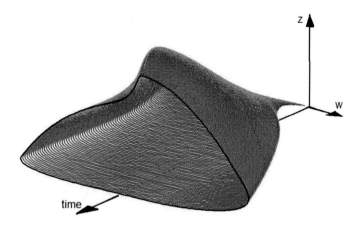

Figure 9.4 3D image of the functional sensitivity set.

$P = 4$, $V_{ref} = 254$ V, $L = 10.1$ Nm, $I = 0.08$ kg m^2, and set point = 1800 rpm.

Contour A was obtained for L and w fluctuating by ±5% of their nominal values, and contour B was obtained with fluctuations of ±8%. The model is quite sensitive to the fluctuations of w. If w is fixed, with the load L fluctuating by ±5%, the boundary of the reachable set is given by contour C. This low sensitivity to L is because of the relatively small value of L and due to the action of the controller.

Fig. 9.4 shows a 3D image of the reachable set. Model parameters for the sets of Fig. 9.3 and Fig. 9.4 correspond to the stable operating region. If the saturation restriction is changed and a stronger controller action is applied, then the model non-linearities are reflected in the images, and the sensitivity set grows. This may affect the robustness of the system. Fig. 9.5 shows the contour of the reachable set for the following parameters: $K = 1.6$, $T_i = 2$ s, $y_m = 700$, $P = 4$, $V_{ref} = 254$ V, $L = 10.1$ Nm, $I = 0.08$ kg m^2, set point = 1800 rpm, and fluctuations in L and w of ±8%.

The reachable set becomes "irregular," approaching the break-down torque and an unstable operating region. In such cases, the DI solver algorithm may reveal some irregularities, caused mainly by the time discretization and imperfections of the numerical integration algorithm.

9.4.1 Comparison with classical sensitivity analysis

Let us run the model with another saturation restriction and greater final time, with the following parameters: $K = 0.5$, $T_i = 2$ s, $y_m = 700$, $P = 4$,

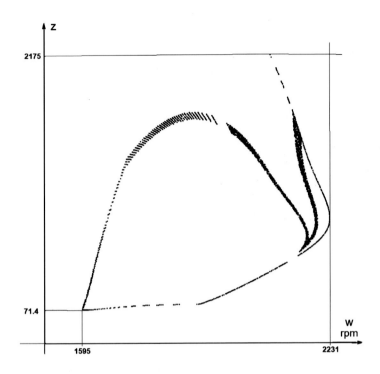

Figure 9.5 Strong non-linearities revealed with greater amplitudes of the disturbances.

$V_{ref} = 254$ V, $L = 10.1$ Nm, $I = 0.08$ kg m^2, set point $= 1800$ rpm, and final time $= 50$.

Fig. 9.6 depicts the contour of the final reachable set. Note the set of points marked with X. The solver displays these points as the boundary points, but, in fact, they belong to the interior of the reachable set. This is due to the fact that the solver is based on the concepts of the maximum principle, which provides necessary and not sufficient conditions for optimality. Anyway, each displayed point belongs to the reachable set.

In Fig. 9.6 the reachable set obtained by "Vensim-like" SA can also be seen. Analysis of this kind is provided by many system dynamics software tools, like Vensim or PowerSim. In this type of analysis, a series of trajectories with randomly changed model parameters (w and L in our case) is calculated. The parameters are different for each trajectory, but constant along each trajectory. It can be seen how different sensitivity sets are provided by the functional sensitivity. Fig. 9.7 shows the projections of the true reachable set and the Vensim analysis onto the time–state plane (a "side view").

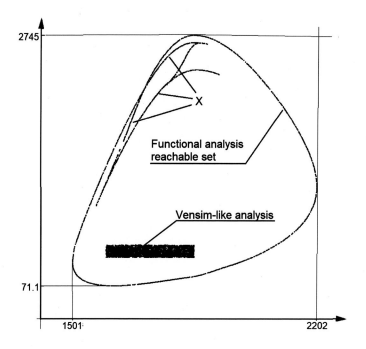

Figure 9.6 Reachable set for time = 50 and Vensim-like sensitivity set.

9.5. Conclusion

Functional sensitivity is closely related to DIs. Here, the non–local functional sensitivity is used. It is defined as the reachable set of the model with uncertain parameters. This approach is deterministic. The disturbances are not treated as random variables. Such deterministic approach seems to be more useful in problems of robust control, when the "worst-case" behavior is important, rather than in probabilistic models.

The velocity control system of induction motors is a highly non–linear dynamic system. The design of such type of control must be done taking into account the robustness of the system. Local stability is not sufficient. The system must respond satisfactorily to small, as well as to big disturbances. The use of the functional sensitivity described in this chapter may help in designing robust control systems. The images of the sensitivity sets produced by the DI solver show the possible extreme deviations of the system trajectories from the desired solutions. The proposed method is dynamic, and permits time-varying disturbances and fluctuations of parameters, unlike in the classical approach to SA.

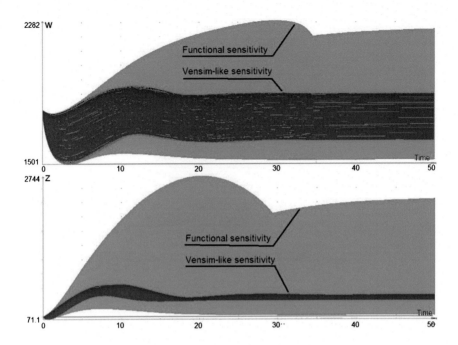

Figure 9.7 Reachable set projections to time-*w* and time-*z* planes.

Further research should be done on the DI solver. The representation of results like the images of the reachable sets should be improved. In the multidimensional case, 2D projections of the point cloud produced by the solver can be shown. It is not an easy task to clearly display multidimensional clouds of points. Animated images and other virtual reality methods may be useful.

References

[1] M. Friendly, D. Dennis, The early origins and development of the scatterplot, Journal of the History of the Behavioral Sciences 41 (2) (2005) 103–130, https://doi.org/10.1002/jhbs.20078.

[2] D.A. Freedman, Statistical Models: Theory and Practice, Cambridge University Press, 2005.

[3] R.D. Cook, S. Weisberg, Criticism and influence analysis in regression, Sociological Methodology 13 (1982) 313–461.

[4] I. Sobol, Sensitivity analysis for non-linear mathematical models, Mathematical Modeling and Computational Experiment 1 (1993) 407–414.

[5] K. Sriyudthsak, H. Uno, R. Gunawan, F. Shiraishi, Using dynamic sensitivities to characterize metabolic reaction systems, Mathematical Biosciences 269 (2015) 153–163.

[6] J.W. Forrester, Industrial Dynamics, Pegasus Communications, Waltham, MA, 1961.

[7] S. Raczynski, Differential inclusion solver, in: Conference Paper: International Conference on Grand Challenges for Modeling and Simulation, The Society for Modeling and Simulation Int., San Antonio, TX, 2002.

[8] L.S. Pontryagin, The Mathematical Theory of Optimal Processes, Wiley Interscience, New York, 1962.

[9] K. Zhou, J.C. Doyle, Essentials of Robust Control, Prentice Hall, ISBN 0-13-525833-2, 1997.

[10] R. Mujal-Rosas, J. Orrit-Prat, General analysis of the three-phase asynchronous motor with spiral sheet rotor: operation, parameters, and characteristic values, IEEE Transactions on Industrial Electronics 58 (5) (2011) 1799–1811, https://doi.org/10.1109/TIE.2010.2051397.

[11] L. Chen, X. Wang, Y. Min, G. Li, L. Wang, J. Qi, Modelling and investigating the impact of asynchronous inertia of induction motor on power system frequency response, International Journal of Electrical Power & Energy Systems 117 (2020), https://doi.org/10.1016/j.ijepes.2019.105708.

[12] A.G. Conrad, Induction motor, AccessScience, 2020, https://doi.org/10.1036/1097-8542.341600.

[13] S. Goolak, O. Gubarevych, E. Yermolenko, M. Slobodyanyuk, O. Gorobchenko, Mathematical modeling of an induction motor for vehicles, Eastern-European Journal of Enterprise Technologies 2 (2(104)) (2020) 25–34, https://doi.org/10.15587/1729-4061.2020.199559.

[14] C.H.B. Apribowo, M.H. Adhiguna, F. Adriyanto, H. Maghfiroh, Performance analysis of three phase induction motor based on ATV12HU15M2 inverter for control system practicum module, Journal of Electrical, Electronic, Information and Communication Technology 2 (1) (2020), https://doi.org/10.20961/jeeict.2.1.41356 (e-ISSN 2715-1263).

[15] J.P. Aubin, A. Cellina, Differential Inclusions, Springer-Verlag, Berlin, ISBN 978-3-642-69514-8, 1984.

[16] D.G. Cacuci, Sensitivity and Uncertainty Analysis, Chapman & Hall/CRC, London, ISBN 1-58488-115-1, 2003.

[17] J. Nearing, Mathematical Tools for Physics, Dover Publications, 2010.

[18] L.D. Elsgolc, Calculus of Variations, ISBN 978-0486457994, 2007.

[19] J.S. Arora, J.B. Cardoso, Variational principle for shape design sensitivity, Aerospace Research Central 30 (2) (2012) 538–547, https://doi.org/10.2514/3.10949.

[20] B.S. Mordukhovich, Sensitivity analysis for generalized variational and hemivariational inequalities, Advances in Analysis (2005) 305–314, https://doi.org/10.1142/9789812701732_0026.

CHAPTER TEN

Uncertainty in public health: epidemics

Abstract

Some of the most popular models of the dynamics of epidemics are discussed, and a brief survey of the literature is provided. Most models are modifications of the *susceptible–infected–removed* model. Simple simulations are carried out. Then, uncertainty is introduced to the models, which leads to the corresponding differential inclusions. The differential inclusion solver is used to calculate the reachable sets and to assess the ranges of the model variables when the uncertain variables are subject to fluctuations.

Keywords

Differential inclusion, Uncertainty, Public health, Reachable set

10.1. Epidemic dynamics

After an epidemic in a human or animal population is detected, an important task is to assess how dangerous it is and which part of society may be infected. So, models of epidemic spread dynamics have been the topic of research for a long time. One of the first classical models of the phenomenon appeared in the late 1920s, in the works of Kermack and Kendrick [7]. In fact, the models that have been developed later on are of a similar type: They try to reflect the dynamics of epidemics using the *system dynamics* (SD) approach, based on *ordinary differential equations* (ODEs). ODE models are normally given in the form of a set of linear or non-linear differential equations, and their properties are widely discussed and improved in many available works. However, ODE or SD models (Forrester [5]) can hardly reflect geospatial factors. The popularity of SD models rose among modelers mainly due to the strange conviction that everything that happens in the real world can be described by differential equations. In fact, this is not exactly true. Some remarks on these issues can be found in Raczynski [13]. This and other deficiencies of ODE and SD models have inspired the development of other modeling and simulation tools, such as object- and agent-based simulation; see Obaidat and Papadimitriou [11] and Perez and Dragicevic [12]. Dargatz and Dragicevic [3] consider an application of an extended *susceptible–infected–removed* model of the spatiotemporal spread

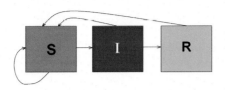

Figure 10.1 The system dynamics scheme of the SIR and SIRS models.

of influenza in Germany. The inhomogeneous mixing of the population is taken into account by the introduction of a network of sub-regions. A multivariate diffusion process is used to describe the model.

Important remark: I appreciate efforts on COVID-19 pandemic simulations that perhaps can result in some useful tools in the future. However, note that the *most serious mistake* in any scientific work is *to look for something that does not exist*. The problem of the existence of solutions is well known in mathematics, but somewhat vague in other fields of scientific research. For new epidemics, caused by new bacteria or viruses, the existence of a valid model may be doubtful. The worst error would be to take a plot of certain past epidemics and then look for a forecast by using the data of the initial period of the disease and some best-fit method to estimate parameters and generate forecasts. Such curves and forecasts may be (and, in fact, were) used by health organizations and governments to take decisions on future actions. This may result in *erroneous disease handling* and in thousands of unnecessary infections and deaths. Anyway, this is also a question of ethics in modeling and simulation.

Continuous SD models result to be useful and can provide, to some extent, important information about disease propagation. The main topic of this book is differential inclusion (DI) modeling, which is closely related to differential equations, but can provide more information about the modeled system.

The basic and most popular is the *susceptible–infected–removed* (SIR) model. Fig. 10.1 shows the SD scheme of the model. Block S represents the number of susceptibles, i.e., the individuals that can be infected. Block I denotes the number of actually infected individuals. R is the number of individuals who previously had the disease and are now immune or dead. A modification of the SIR model named SIRS includes also a feedback from block R to S, as shown in Fig. 10.1. The equations of the SIR model are as follows (r and a are constants):

$$\begin{cases} \dfrac{dS}{dt} = -rS(t)I(t), & \text{(A)} \\[2mm] \dfrac{dI}{dt} = rS(t)I(t) - aI(t), & \text{(B)} \\[2mm] \dfrac{dR}{dt} = aI(t). & \text{(C)} \end{cases} \qquad (10.1)$$

The coefficient r is the disease contraction rate and a represents the mean recovery/death rate. Consult Misic and Santarelli [9] for SIR model simulation.

Eq. (10.1A) tells us that the speed by which the number of susceptibles decreases is proportional to the product SI, which is proportional to the number of possible contacts between infected and susceptibles. In some versions the product SI is divided by the total population. Here, we assume that this factor is included in the coefficient r and that the changes of the whole population are slow. Eq. (10.1B) describes the rate of change of the number of infected individuals, which is equal to SI multiplied by the disease contraction rate minus a term proportional to I (the individuals that recover or die). Finally, Eq. (10.1C) defines the rate of increase of the number of removed individuals. Note that in the SIR model, the variables S and I do not depend explicitly on R. So, if we do not need the solution for $R(t)$, we can use only Eqs. (10.1A) and (10.1B). The SIR model is perhaps too simplified. It has also been criticized for its high sensitivity to variations of the parameter r.

If we suppose that some of the recovered people may be infected again, we can use a modification of the SIR model, named SIRS (the prime mark stands for time differentiation):

$$\begin{cases} S' = -rS(t)I(t) + m(N - S(t)) + fR(t), \\ I' = rS(t)I(t) - (a + m)I(t), \\ R' = aI(t)(C) - (m + f)R(t), \end{cases} \qquad (10.2)$$

where a, r, m, and f are constants. Note that the term $fR(T)$ represents the number of recovered individuals that join the S group. Coefficient f tells which part of R will be incorporated again into group S. N is the total population.

The SEIS model is a modification of the SIR model, where there is an additional group E that represents the exposed population during the latent period of the disease. The equations are as follows:

$$
\begin{cases}
S' = B - rS(t)I(t) - mS(t) - rS(t)I(t) + aI(t), \\
E' = rS(t)I(t) - (e + m)E(t), \\
I' = eE(t) - (a + m)I(t),
\end{cases}
\tag{10.3}
$$

where a, r, m, e, and B are constants.

There are a lot of modifications of the above models. For epidemics with larger duration, the birth-and-death process is added. Passive immunity is taken into account in the MSIR model, where it is supposed that some individuals are born with immunity. The MSEIR model has the scheme $M \to S \to E \to I \to R$, where M is the number of passively immune individuals. Supposing that the immunity in group R is temporary, we obtain the MSEIRS model $M \to S \to E \to I \to R \to S$.

In the article of Ng et al. [10] we can find a description of a model of a double epidemic. Two superimposed epidemics are considered using a modification of the SIR model. The problem is focused on the Hong Kong SARS epidemic in 2003, caused by two different viruses. The resulting model has the form of a system of six differential equations of the first order.

Some more complicated models of epidemic dynamics can be found in Gebreyesus and Chang [6]. They propose a multicompartment model that takes into account interactions between humans and animals or between different species of animals. It is a multivariable state-space model that reflects the phenomena of some diseases transmitted from animal to human, such as Ebola, MERS, bird flu, and tuberculosis. The basic model used in that article is a modification of the SIR model. The main concept is to define a number of clusters in the populations, where the epidemics are governed by SIR equations, interacting with each other. The model includes spatial issues, introducing regions like urban and rural. For models of recurrent epidemics, see David [4].

The above models are deterministic. In order to manage the uncertain elements, stochastic elements have been introduced in disease spread models. Various types of stochastic models are discussed by Allen [1], who considers some modifications of the basic SIR model. The mathematical tools include Markov chains and stochastic differential equations. From these models we can obtain the probability distributions for final size of the S, I, and R groups, for the disease extinction, disease duration, and other parameters. In the Allen model, S, I, and R are treated as discrete random variables. A detailed mathematical background is given, where the obtained equations describe the probabilities rather than the instant values of the variables. MATLAB® code is given as well. See also Matis and Kiffe [8]. An overview of various modeling techniques, deterministic and stochastic, can be found in Daley and Gani [2].

As stated in previous chapters, our approach to uncertainty is quite different. The proposed approach, based on DIs, is deterministic. The variables treated as random in the stochastic models are supposed to be uncertain, but not necessarily random. We do not obtain any probabilistic results, managing rather the possible range of uncertainty given as the resulting attainable sets.

10.2. Modeling tool

The main mathematical tool we use to calculate models' reachable sets is the *DI*. The numerical implementation of DI models is achieved by the DI solver. The background of DIs is given in Chapter 1. Here, let us recall only the main concepts. A differential equation, in the canonical form, has a (scalar or vector) right-hand side, which is a function of the independent variable and the state variables. The right-hand side of a DI is a set-valued function:

$$\begin{cases} x' \in F(x, t), \\ x(0) = X_0, \ x \in R^n, \ F \subset R^n, \end{cases} \tag{10.4}$$

where F is a set-valued function and $F : R^n \times R \to P(R^n)$ (subsets of R^n).

A function $x(t)$ that satisfies (10.4) a.e. over a given time interval is called a *trajectory* of the DI. Sometimes such function is called a solution to the DI. However, in this book we treat such functions as trajectories, and the *reachable set* of the DI as the solution. This is because the reachable set is a natural generalization of a single solution to a differential equation. Recall that the reachable (or attainable) set is the union of the graphs of all trajectories of the DI. With the necessary regularity assumptions (see Chapter 1), the reachable set exists and is unique. To avoid a conflict of terminology with other sources, we will not use the term solution, talking rather about reachable sets. See Chapter 1 for more detail.

The *DI solver* is used to calculate the shape of the reachable set. See Chapter 2 for a detailed description of the solver algorithm. Only recall that the main concept of the solver is to scan the boundary, and not the interior, of the reachable set by trajectories obtained using the methods from optimal control theory. In Chapter 2 it is pointed out that simple shooting scanning the interior of the set using randomly generated trajectories may lead to poor assessments and wrong results.

The way we obtain the DI for a given ODE model is based on the fact that the DIs are closely connected to control systems. We consider an ODE model with uncertain and variable parameter, supposing that we have no or poor information about the probabilistic properties of the parameter(s), such as the probability distributions and confidence intervals, but we can assess the intervals where the parameters must belong. So, if the parameters scan their corresponding ranges, the (vectorial) value of the right-hand side of model equations scans a set, which is exactly the set F of the inclusion (10.4). Next, we analyze the obtained DI using the DI solver.

10.3. Examples of reachable sets

As in previous chapters, the model itself is not the main topic of this book. Our aim is to show an alternative approach to uncertainty, rather than developing new models. To carry out some simulation experiments we use the basic SIR and SIRS models, adding uncertainty to some of its parameters. Similar simulations can be carried out with any other model.

10.3.1 The SIR model

The model equations for the SIR model are as follows:

$$\begin{cases} S' = -r[1 + 0.05u_1(t)]S(t)I(t), \\ I' = r[1 + 0.05u_1(t)]S(t)I(t) - a[1 + 0.05u_2(t)]I(t), \\ R' = a[1 + 0.05u_2(t)]I(t), \end{cases} \quad (10.5)$$

where $u_1(t)$ and $u_2(t)$ are functions of time and belong to the interval $[-1, 1]$. In other words, we assume that the coefficients r (disease contraction rate) and a (mean recovery/death rate) have uncertain values and may change in time, within the interval $\pm 5\%$ of their default values. When u_1 and u_2 scan all their permissible values, then the right-hand side of (10.5) scans a 3D set, being the right-hand side of the corresponding DI.

To see how the model behaves, we first run a simple simulation with u_1 and u_2 constant and equal to zero. This converts the model (10.5) into the basic SIR model. Model parameters are as follows: disease contraction rate

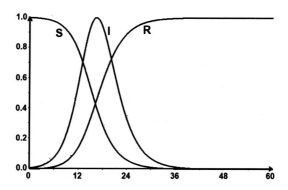

Figure 10.2 Simple simulation of the SIR model.

$r = 0.00001$, mean recovery/death rate $a = 0.6$, $S(0) = 100,000$, $I(0) = 50$, $R(0) = 0$, and final simulation time $= 60$.

Fig. 10.2 shows the results of the simulation. The three curves are normalized into the interval 0–1. The ranges of the variables are: S between 33,300 and 100,000, I between 0 and 9413, and R between 0 and 67,700, approximately.

Now, we allow the parameter r to be uncertain; it can change between 0.0000075 and 0.0000125. Let us calculate the reachable set for the model variables. Note that the reachable set at the final simulation time is not very interesting because the variables reach the final steady state. So, we will rather look for the reachable set in shorter simulation time, equal to 20. The reachable set is shown in Fig. 10.3.

Running the model with both r and a uncertain, we obtain a larger reachable set, as shown in Fig. 10.4. From the resulting images of reachable sets, we can learn what are the limits for possible variable changes. Note that those are not global maximal and minimal values, but rather the limits for the time instant under consideration. Of course, similar images can be obtained for any other, user-defined time instant.

In Fig. 10.5 we can see the 3D image of the reachable set with both r and a uncertain. The coordinates are the time, the number of susceptibles S, and the number of infected individuals I. The origin of the coordinate axes is fixed at 0, 100,000, and 50, respectively. Fig. 10.6 shows the projection of some randomly selected (but not random) trajectories for the case with r and a uncertain onto the time-infected plane. These are projections of the trajectories that scan the boundary of the 3D reachable set. It can be seen

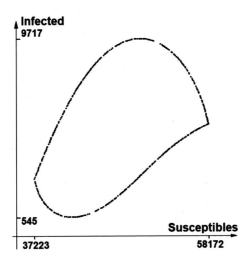

Figure 10.3 Time section of the reachable set, with parameter *r* uncertain, time = 20.

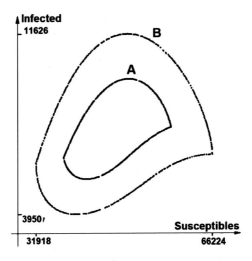

Figure 10.4 Reachable set at time = 20. (A) Uncertain *r*. (B) Uncertain *r* and *a*.

that the number of infected individuals changes considerably even while the range of uncertain parameters is ±5%.

Similar simulations for the final time equal to 40 are shown in Fig. 10.7, which shows the reachable set time section, and Fig. 10.8, where we can see some boundary scanning trajectories.

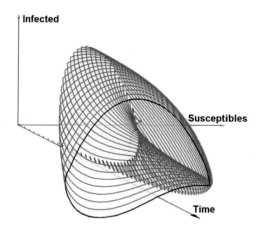

Figure 10.5 3D image of the reachable set, with parameters r and a uncertain and time $= 20$.

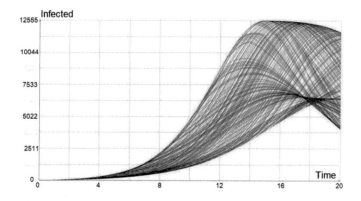

Figure 10.6 Some boundary scanning trajectories, with r uncertain.

To see another possible scenario of epidemic spread, suppose that at time equal to 10 the health services take some measures, for example implementation of a new, effective vaccine. This reflects on the coefficient a (recovery/death rate), which becomes two times greater.

Fig. 10.9 depicts the simple simulation. At time $= 10$ the recovery accelerates, and the maximal number of infected individuals decreases to 2585. It should be noted that in this case we suppose the vaccine is "ideal" (with efficiency 100%) and that it is applied immediately to the whole population. In reality, certain inertial delay should be added.

In Fig. 10.10 we can see some boundary scanning trajectories for this case. Here, we have the same uncertainty as before, i.e., both controls r and

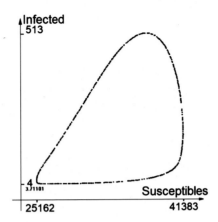

Figure 10.7 Shape of the boundary of the reachable set for time = 40.

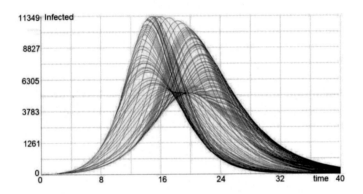

Figure 10.8 Some boundary scanning trajectories. Final time = 40.

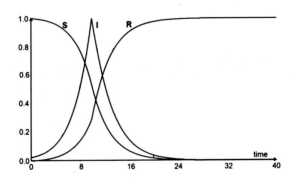

Figure 10.9 Simple simulation with mean recovery/death rate increased at time = 10. Final time = 40.

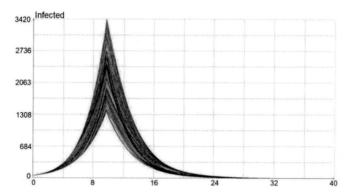

Figure 10.10 Some boundary scanning trajectories with expected recovery/death rate increased at time $= 10$.

a change between ±5%. Compared with Fig. 10.8, it can be seen that the maximal number of infected individuals decreases from 12,555 to 3420.

10.3.2 The SIRS model

To see another model version, we use the SIRS model (10.2). In this model we assume that recovered people may be infected again. The equations with the two controls inserted are as follows:

$$\begin{cases} S' = -r(1 + 0.05u_1(t))S(t)I(t) + m(N - S(t)) + fR(t), \\ I' = r(1 + 0.05u_1(t))S(t)I(t) - a[(1 + 0.05u_2(t)) + mI(t)], \qquad (10.6) \\ R' = a(1 + 0.05u_2(t))I(t) - (m - f)R(t). \end{cases}$$

The controls are used in the same way as in the previous SIR model, supposing that the parameters r and a of the SIR model are uncertain variables.

To compare the models, we use similar parameters: disease contraction rate $r = 0.00001$, mean recovery/death rate $a = 0.6$, $S(0) = 100{,}000$, $I(0) = 50$, $R(0) = 0$, and final simulation time $= 60$.

In addition, we define $m = 0.07, f = 0.04$, and $N = 100{,}000$.

A simple simulation of this model with final time equal to 60 is shown in Fig. 10.11.

The time section of the reachable set for this model is shown in Fig. 10.12, and some boundary scanning trajectories are shown in Fig. 10.13. The section time is equal to 40, and the uncertainty ranges for r and a are the same as for the SIR model (±5%). Now, all three equa-

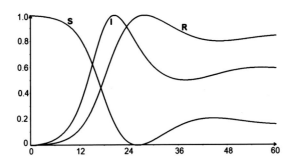

Figure 10.11 Simulation of the SIRS model. Final time $= 60$.

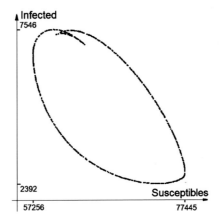

Figure 10.12 Time section of the reachable set for the SIRS model with uncertainty. Final time $= 40$.

tions interact with each other, so the simulation of the first two of them cannot be done as in a 2D model. The shape is more complicated, and the boundary of the reachable set may fold with itself.

10.4. Conclusion

A general conclusion from the experiments shown in this chapter is that even with small uncertainty of model parameters, the uncertainty of the resulting model state may be quite big. This means that the models hardly provide exact predictions, even for short time horizons.

DIs may provide important additional information about the behavior of SD models. In fields such as public health, problem management,

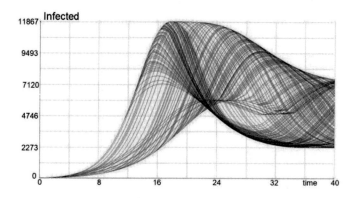

Figure 10.13 SIRS model. Some boundary scanning trajectories.

marketing, or economy, the models are charged with a great amount of uncertainty. Treating uncertain factors as stochastic variables may be useful, but it is not always possible or conceptually correct. Not everything that is uncertain can be represented by random variables. To manage probabilities, we must be able to get or measure properties such as basic statistics, and we must be sure that probability distributions and other statistics of considered variables exist at all. In public health problems, for example, the decisions of (external) health authorities can be uncertain, but not random. DI models can provide results such as the attainable sets, which are deterministic objects that are useful when analyzing the dynamics of the model.

References

[1] L.J.S. Allen, E.J. Allen, An Introduction to Stochastic Epidemic Models, Springer-Verlag, Berlin, Heidelberg, 1945.

[2] D.J. Daley, J. Gani, Epidemic Modelling: An Introduction, Cambridge University Press, ISBN 978-0-521-01467-0, 2001.

[3] C. Dargatz, S. Dragicevic, A Diffusion Approximation for an Epidemic Model, Collaborative Research Center 386, 2006.

[4] J.D. David, A light introduction to modelling recurrent epidemics, in: Mathematical Epidemiology, in: Lecture Notes in Mathematics, 1945.

[5] J.W. Forrester, Industrial Dynamics, Pegasus Communications, Waltham, MA, 1961.

[6] K.D. Gebreyesus, C.H. Chang, Infectious diseases dynamics and complexity: multi-compartment and multivariate state-space modeling, in: Conference Paper: Proceedings of the World Congress on Engineering and Computer Science 2015, 2, San Francisco, ISBN 978-988-14047-2-5, 2015.

[7] W.O. Kermack, A.G. McKendrick, Contributions to the mathematical theory of epidemics, Part I, Proceedings of the Royal Society of Edinburgh. Section A. Mathematics 115 (772) (1927) 700–721.

[8] J.H. Matis, T.R. Kiffe, Stochastic Population Models. A Compartmental Perspective, Springer, 2000.

 [9] L. Misici, F. Santarelli, Epidemic propagation: an automaton model as the continuous SIR model, Applied Mathematics 4 (2013) 84–89.

[10] T.W. Ng, G. Turinici, A. Danchin, A double epidemic model for the SARS propagation, Journal of Negative Results in Biomedicine 3 (19) (2003), https://doi.org/10.1186/1471-2334-3-19.

[11] M.S. Obaidat, G.I. Papadimitriou, Applied System Simulation Methodologies and Applications, Springer, ISBN 978-1-4613-4843-6, 2003.

[12] L. Perez, S. Dragicevic, An agent-based approach for modeling dynamics of contagious disease spread, BioMed Central 8 (50) (2009), https://doi.org/10.1186/1476-072X-8-50.

[13] S. Raczynski, Discrete event approach to the classical system dynamics, in: Conference Paper: Huntsville Simulation Conference Huntsville Alabama, The Society for Modeling and Simulation, 2009.

CHAPTER ELEVEN

Uncertain future: a trip

Abstract

There are many situations where people intend to predict or assess events from the future. In the field of modeling and computer simulation we can find various "predictor" algorithms that, to some extent, provide information about the possible future. In this chapter, we discuss models with the "ideal predictor," where the model actions depend on (exact) information from the future. As we cannot obtain such information, we suppose that the future is uncertain and consider the output from the ideal predictor as a set of possible values. This leads to a model expressed in the form of a differential inclusion.

The content of this chapter is not science fiction, though some fictitious elements can be found here. We discuss an abstract model with feedback from the future (ideal predictor).

Keywords

Differential inclusion, Uncertainty, Future uncertainty, Reachable set

11.1. Uncertain future

We discuss a dynamic model with positive time shift (ideal predictor). The element with unknown future state is replaced with a set that represents the uncertainty of the future. This leads to a differential inclusion (DI). The DI solver provides the solution to this inclusion, in the form of a reachable set in the time–state space. Then, an iterative process is proposed that converges to a single trajectory, being the solution to the original problem with the positive time shift. Some examples of linear and non-linear models with ideal predictor are given. These models with positive time shift are different from the known predictor algorithms. We use the DI solver to deal with the problem, instead of differential equations.

If we suppose that our world is causal, then we should discard the possibility of traveling to the future, going back, and making decisions or acting according to the information taken from the future. Many paradoxes arise from time traveling. Let us recall only one, perhaps the most illustrative. Imagine that young Beethoven travels to the future and goes to a concert where his fifth symphony is played. Then, he returns to the original time. As he has perfect musical memory, he converts the music he heard into musical score and publishes it. Thus, the fifth symphony has never been composed.

This chapter is not science fiction. We discuss just a possibility of simulating an ideal predictor (which, in fact, contradicts the principle of causality). Note that we do not deal with any known prediction algorithms used in decision making processes in marketing, economy, stock market trading, control, and other fields. Such predictors provide an approximate, perhaps more probable scenario for future system behavior, but they are not ideal predictors. By "ideal predictor" we mean an object that gives us exact information about the future system state.

Both the literature and the Web are full of articles and other publications on time traveling. Most of them belong to the field of science fiction, so we will not discuss them. From some more serious publications, let us mention Nahin [3], where we can find remarks on time concepts in quantum physics and more references.

In the field of system dynamics and control theory, there are many publications that deal with time-delayed systems. These considerations are more realistic because we can, to some extent, have information about the history of the system under consideration. This information may be supposed to be exact or charged with some uncertainty. On the other hand, the future system states are more uncertain, and the uncertainty grows with the time-distance to the future system state.

If we restrict our problem to continuous ordinary differential equation (ODE) models, then we could consider the equation

$$\frac{dx}{dt} = f(x(t), x(t+r), t). \tag{11.1}$$

Here, x is a point in the 1D or multidimensional state space, r is the time shift (negative for time-delay systems), and f is a vector-valued function. A linear version of the time shift system is frequently given by the following equation:

$$\frac{dx}{dt} = \mathbf{A}x(t) + \mathbf{B}x(t+r), \tag{11.2}$$

where $\mathbf{A}$ and $\mathbf{B}$ are matrices. If r is less than or equal to zero, there is no problem with simulating the above models, with given initial conditions for x. If no analytical solution can be found, we can apply any of the known numerical methods for ODEs. The presence of the delayed argument requires the past model trajectory to be stored. Having such a history file, we can retrieve the data necessary to advance with the solution in time. If the time shift r is positive, integration of the model trajectory is difficult.

The second argument of the function f of (11.1) is unknown and uncertain. Now, let us treat this argument not as a point in the state space, but as a set V, as follows:

$$\frac{dx}{dt} = f(x(t), V(t), t), \qquad (11.3)$$

where V is a set, maybe changing in time. This set represents the uncertain future. This way, our model (11.1) takes the form of a DI, instead of a differential equation:

$$\frac{dx}{dt} \in F(x(t), t). \qquad (11.4)$$

Here, F is a set-valued function, defined as follows:

$$F(t, x) = \{z : z = f(x, u, t), \, u \in V(t)\}. \qquad (11.5)$$

Consequently, our problem is to solve a DI, instead of differential equation. In practice, we can assess the possible limits of V. The solution to a DI is a set, called *reachable* or *attainable set*, discussed in Chapter 1. Observe that the equations of the original problem (11.1) include undefined elements. In turn, in the corresponding DI (11.4) nothing is undefined. Instead of uncertain parameters, we have a well-defined set. DIs may be used as an alternative to stochastic and probabilistic approaches to uncertainty. Instead of stochastic, we can use *tychastic* variables, whose values are uncertain, but not random; see Aubin [1].

Fig. 11.1 shows a rough explanation of our problem. It is supposed that we have exact information about the present model state x (it also could be questionable). The future states are uncertain (region X). In the figure, some uncertainty of the past states is also assumed.

When using DIs, the problem with time delay is not very different from the problem of the ideal predictor. In both cases, we need to solve the corresponding DI. In this section, we consider a somewhat abstract problem of systems with positive time shift, and we do not care about the causality principle. We are just looking for a method to solve differential equations with positive time shift. Some examples are given further on.

After solving the inclusion (11.4) we obtain a reachable set that may be smaller than the set X. In the following, an iterative algorithm is discussed that may provide the solution to the model with positive time shift.

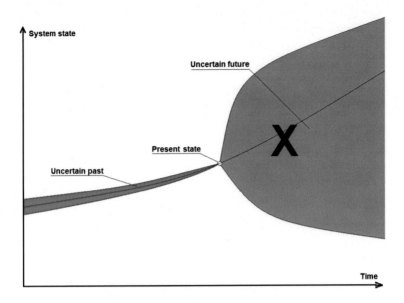

Figure 11.1 System trajectory, uncertain past and future.

11.2. Differential inclusion solver

A discussion of DIs was provided in Chapter 2. Here, let us recall the applications of the DI solver.

The basic version of the DI solver is not new. It was published in 2002; see [5]. In Chapter 2, a new version of the solver and some new applications to the functional sensitivity are described. The DI solver and the present problem statement should not be confused with the DI method used in optimal control problems.

To avoid repetition, here we will not discuss the algorithm of the DI solver in detail. Recall that the DI solver generates a series of DI trajectories that scan the boundary, and not the interior, of the reachable set. The DI is derived from the model state equation of the following equivalent control system, which includes the vector of uncertain parameters $u = (u_1, u_2, ...u_m)$:

$$\frac{dx}{dt} = f(x(t), u(t), t), \ t \in I = [0, T], \ T > 0. \tag{11.6}$$

Vector u belongs to a given set of restrictions $C(x, t)$. When u scans the interior of the set $C(x, t)$, the right-hand side of (11.6) scans a set F, being the right-hand side of the corresponding DI (11.4).

The solver algorithm uses some results from optimal control theory (Pontryagin [4]). From Pontryagin's maximum principle it is known that each model trajectory that reaches a point on the boundary of the reachable set at the final simulation time must belong to the boundary of this set for all earlier time instants. Consult Pontryagin [4] and Lee and Markus [2]. Moreover, a boundary scanning trajectory must satisfy the Jacobi–Hamilton equations (a necessary condition). After integrating a sufficient number of such trajectories, we can see the shape of the boundary of the reachable set. Consult Chapter 2 for details of the solver algorithm.

Someone might suppose that generating a number of trajectories randomly, we can cover the inside of the reachable set with sufficient density, and then estimate its shape. Unfortunately, this is not true, even if we select only points from the boundary of the set F. Simulation experiments show that even in very simple cases the set of trajectories provided by primitive shooting (using any density function) is concentrated in some small region inside the reachable set and does not approach its boundary. Anyway, it is an error to explore the interior of the reachable set. The DI solver used here explores the boundary of the set, and not the interior.

11.3. The ideal predictor: feedback from the future

Suppose that the present state of the system is known (see Fig. 11.1), and the future states are uncertain, with given (maybe big) permissible set. Our model is given in the form of a DI (Eq. (11.4)). Using the DI solver, we can solve the inclusion and obtain the reachable set for the future time instants. The resulting reachable set can be greater or smaller than the estimate of the limits of uncertain future states. As the initial estimate is normally given as a big set, it is probable that the reachable set we obtain is smaller than the initial uncertainty estimate. In many practical applications this is the case because many physical systems have some limiting elements like saturation and delimiters. Now, we replace the original future uncertainty estimate with the reachable set provided by the last run of the solver, and solve the inclusion again. Repeating these steps, we get an iterative process that can converge or not to a single function $x(t)$. If the process converges, then we obtain a very narrow reachable set, which is an estimate of the solution to the original problem with ideal predictor.

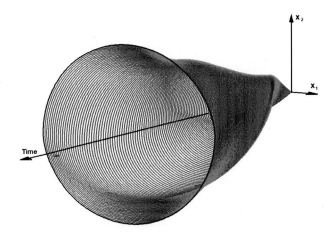

Figure 11.2 3D image of the reachable set for model (11.7).

11.3.1 Example 1: a linear model

Consider the following model:

$$\begin{cases} x_1' = x_2 - 0.07x_1(t+0.6), \\ x_2' = 10 - x_1(t) - 0.35x_2(t+0.6), \end{cases} \tag{11.7}$$

with initial condition $x_1(0) = x_2(0) = 0$ and final simulation time equal to 6. Fig. 11.2 shows a 3D image of the boundary surface of the reachable set for model (11.7). In this run of the solver, the initial uncertainty of the state variables was defined as a permissible rectangle that delimits both variables to the interval ± 100. The range of the final state resulted to be even greater, with $-165 < x_1 < 165$ and $-161 < x_2 < 156$, approximately. During the calculation process, the actual ranges for the two variables are stored. In the next iteration, the stored range values are used to calculate the next approximation of the reachable set. In this case, this iterative process produces a sequence of reachable sets that converge the solution to the original time shift problem (11.1).

In Fig. 11.3 we can see a comparison of the consecutive reachable sets for the final time equal to 6 (Fig. 11.3A) and 5 (Fig. 11.3B). Note that for the final interval $[t - r, t]$ we have no range parameters stored for $x(t + r)$, so the set V is taken equal to that of the first iteration (± 100 rectangle). This is why the contour of the final reachable set is always finite and does not converge to one point. The first contour (iteration 0) looks somewhat different from the next contours because in the first iteration the set V

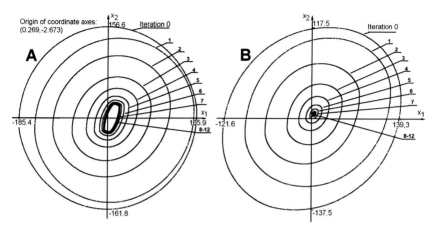

Figure 11.3 Comparison of the contours of the reachable set for (A) final time instant and (B) time $= 5$, with consecutive iterations.

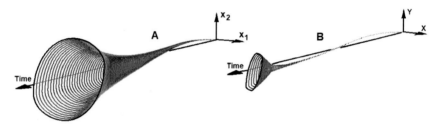

Figure 11.4 Comparison of reachable sets from (A) iteration 5 and (B) iteration 10.

is equal for all time steps, and in the consecutive iterations it is different for each time step (stored before). From Fig. 11.3B, we can see that the convergence of the iteration process is quite good.

Fig. 11.4 shows a comparison of the reachable sets for iterations 5 and 10. The images have the same scale as Fig. 11.3. The convergence of the process is clearly seen.

11.3.2 Example 2: a non-linear model

Consider the following system equations:

$$\begin{cases} x'_1 = x_2 + 0.1x_1(t+r), \\ x'_2 = 20 - x_1(t) - 0.2x_2(t) - 0.024\,x_2^2(t). \end{cases} \tag{11.8}$$

Here, $r = 0.35$ and the final simulation time is equal to 5. See Figs. 11.5 and 11.6.

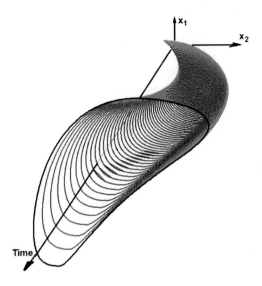

Figure 11.5 Reachable set for model (11.8), with uncertain future state in $[-100, 100]$.

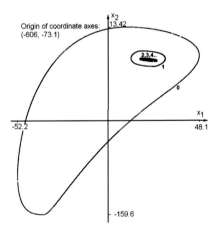

Figure 11.6 The contours of the reachable set for model (11.8), with consecutive iterations.

11.3.3 Example 3: a control system

Consider the control system of Fig. 11.7. The controlled process may be interpreted as a thermal object with simplified transfer function of second order. The "time shift" block can represent time delay or time advance. Supposing a time delay we obtain a typical academic control problem described in elementary books on automatic control. Now suppose that

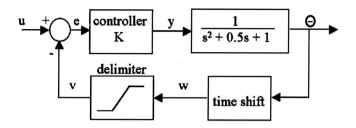

Figure 11.7 A control system with ideal predictor.

instead of time delay we have an "ideal predictor," the "feedback from the future."

Here, Θ is the process output that may be, for example, the temperature to be controlled; u is the set point (desired temperature), K is the controller gain, and y is the controller output signal. The actuator delimiter is described by the function $v = g(w)$. Let us denote $x_1 = \Theta$ and $x_2 = d\Theta/dt$. The output signal of the time shift element is equal to $x_1(t + r)$, where t is the time and r is a positive constant.

With this notation, the state equations for this system are as follows (a simple saturation function is used):

$$\begin{cases} \dfrac{d^2\theta}{dt^2} + 0.5\dfrac{d\theta}{dt} + \theta(t) = y, \\ y = Ke, \quad e = u - v, \\ w = \theta(t + r), \\ v = w \text{ for } w < 2 \text{ and } w > 0, \\ v = 2 \text{ for } w \geq 2, \\ v = 0 \text{ for } w \leq 0, \\ u = \text{const} = 1. \end{cases} \quad (11.9)$$

Note that the set point is equal to one, so the desired operation point for the model output x is also equal to one. Consequently, the delimiter works in the range 0–2, symmetric with respect to this operation point. Other model parameters are as follows: $K = 2$, $r = 0.5$, and final simulation time equal to 8. Thus, we have

$$\begin{cases} \dfrac{dx_1}{dt} = x_2(t), \\ \dfrac{dx_2}{dt} = K(1 - v) - 0.5x_2(t) - x_1(t). \end{cases} \quad (11.10)$$

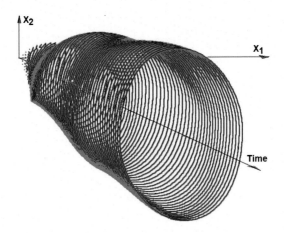

Figure 11.8 Reachable set for the initial future uncertainty.

Due to the assumptions of Pontryagin's maximum principle, the right-hand sides of (11.10) should be continuously differentiable. From the simulation experiments it follows that the solver works also for systems with (not necessarily continuously) differentiable right-hand sides of the state equations.

In Eqs. (11.9) and (11.10) the value of $x(t + r)$ and, consequently, the value of v are uncertain (uncertain future). Replacing w with the set of all its permitted values, we get a DI. Here w is the "control" parameter that parametrizes the set of the right-hand side of the DI. After introducing the above equations to the DI solver, we obtain the following results. Fig. 11.8 shows the system's reachable set for the initial uncertainty range. The resulting limits of w are stored for all time steps. In the next iteration, these limits are used, and so on. In this case the convergence of the iterative process is also quite good. After several iterations we obtain an approximation of the trajectory that satisfies Eq. (11.10). Figs. 11.9 and 11.10 show the shape of the reachable set for iterations 14 and 40, respectively.

11.4. Conclusion

In the above experiments and in other similar situations the algorithm converges quite well. However, there are situations when it does not. The convergence is out of the scope of this chapter and could be a good subject for more theoretical considerations. Note that this is not the problem of the convergence of the DI solver. Here, the solver is used as a part (one

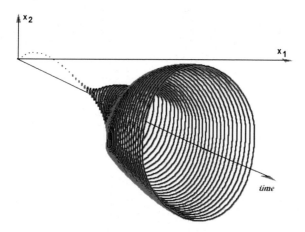

Figure 11.9 Reachable set after iteration 14.

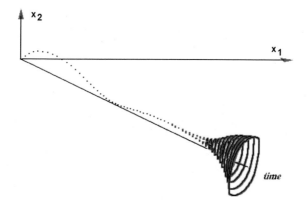

Figure 11.10 Reachable set after iteration 40.

step) of a bigger algorithm. If convergence occurs, we obtain one maybe unique trajectory that satisfies the general rules of movement of our system with the ideal predictor. This means that if the procedure converges to one trajectory, the system with "traveling to the future" element may be stable. But, this is a rather abstract and philosophical question, out of the scope of this book.

It can be shown that a small external disturbance added to the feedback signal affects the convergence considerably. It can also be seen that the stability of the model itself is necessary for the convergence. Running the same model with, for example, $K = 5$, the algorithm does not converge to any single trajectory. As for the disturbances in control systems, if we treat them as uncertainty instead of stochastic functions, we again get a DI. Solving such DI we obtain the range for the output variables. This may be useful in problems such as system safety and robustness.

References

[1] J.P. Aubin, A. Cellina, Differential Inclusions, Springer-Verlag, Berlin, 1984.
[2] E.B. Lee, L. Markus, Foundations of Optimal Control Theory, Wiley, New York, ISBN 978-0898748079, 1967.
[3] P.J. Nahin, Time Machines: Time Travel in Physics, Metaphysics, and Science Fiction, AIP Press, Springer, ISBN 0-387-98571-9, 1998.
[4] L.S. Pontryagin, V.G. Boltyanskii, R.V. Gamkrelidze, E.F. Mishchenko, The Mathematical Theory of Optimal Processes, Interscience, ISBN 2-88124-077-1, 1962.
[5] S. Raczynski, Differential inclusion solver, in: Conference Paper: International Conference on Grand Challenges for Modeling and Simulation, SCS, San Antonio, TX, 2002.

Index

Printed in the United States
by Baker & Taylor Publisher Services